Ecoregions

Robert G. Bailey

Ecoregions

The Ecosystem Geography
of the Oceans and Continents

Second Edition

Robert G. Bailey
Rocky Mountain Research Station
USDA Forest Service
Fort Collins, CO
USA

ISBN 978-1-4939-3706-6   ISBN 978-1-4939-0524-9 (eBook)
DOI 10.1007/978-1-4939-0524-9
Springer New York Heidelberg Dordrecht London

Cover Illustration: Temperate semi-desert in Wyoming, USA. Vintage postcard.

Printed on acid-free paper

Springer is part of Springer Science+Business Media (www.springer.com)

Dedicated to the memory of my son
Matthew Gale Bailey (1968–1998)

Preface to the Second Edition

The first edition of this book (1998) classified and characterized the regional-scale ecosystem units (ecoregions) of the Earth as shown on a map that I developed with the encouragement of several international organizations. In addition to the descriptive account, my primary goal was to suggest explanations of the mechanisms that act to produce the world pattern of ecoregion distribution and to consider some of the implications for land use. I included ocean types since understanding land regions depends on understanding ocean systems.

The Chief of the US Forest Service, Mike Dombeck, distributed this book to Forest Service field offices, Washington Office staff directors, heads of other agencies, Secretaries of Agriculture and Interior, leaders of professional societies and conservation organizations, Vice President Gore, and select members of Congress. He wrote in his transmittal letter, "The Forest Service is beginning to look beyond national forest boundaries and, based on Bailey's work, expand its horizons to view forests from a larger ecoregion-based perspective."

The increasing importance of ecoregions is confirmed by the fact that much planning, research, and management efforts by the Forest Service, The Nature Conservancy, World Wildlife Fund, and other organizations are taking place now within the framework of ecoregions. For example, in 1993, as part of the National Framework of Ecological Units, the US Forest Service adopted ecoregions for use in ecosystem management.

Over the last 14 years since the book was first published, a number of studies have greatly contributed to a better understanding of the Earth's ecoregions. This second edition is a completely updated and expanded version. The main purpose of the revision is to incorporate the latest factual information and the newest geographic ideas. However, it was felt that the book would benefit from new sections that address how ecoregions are changing under the relentless influence of humans (such as modification of fire regimes and the introduction of invasive species) and climate change, use of ecoregional patterns to transfer research results and select sites for detecting climate change effects on ecosystem distribution, use of ecoregional patterns to design monitoring networks and sustainable landscapes, and how the system used in this book compares with other approaches.

This book is intended for several audiences. In addition to environmental planners and decision-makers, it should be particularly useful to those involved with monitoring global change. For example, worldwide monitoring of agricultural and other natural-resource ecosystems is needed to assess the effects of possible climate changes and/or air pollution on our global resource base. Monitoring of all sites is neither possible nor desirable for large areas, and so a means of choice has to be devised and implemented. This is where ecoregions come in. Ecoregion maps show the Earth's land areas subdivided into regions based on large patterns of ecosystems. These regions define large areas within which local ecosystems recur in a predictable pattern. By observing the behavior of the different systems within a region, it is possible to predict the behavior of an unvisited one. Hence the maps can be used to spatially extend data obtained from limited sample sites. The results of observations at representative sample sites from each region are potentially useful in detecting and monitoring global change effects.

Content of this revised edition borrows heavily from my previous works specifically because the repetition provides the information framework necessary to support the new material.

Once again I would like to thank many people who made the completion of this book possible: Nancy Maysmith for re-creating many of the first edition diagrams and drawing the several new ones; Michael Wilson, Program Manager for Inventory, Monitoring, and Analysis at the Rocky Mountain Research Station, for his support. I appreciate the helpful criticism of several reviewers of the first edition; but I should mention especially Fred Smeins, Jeffrey Walck, Philip Keating, Gary Griggs, Andrew Millington, Frank Kuserk, David McNeil, and David Scarnecchia. James Bockheim and Paul Jepson gave numerous helpful suggestions on the content of this new edition. Special thanks to Kevin Cook, who did his best to make the book readable. As always, it has been a pleasure to work with Janet Slobodien at Springer in translating this work to print.

Fort Collins, CO   Robert G. Bailey
December 2013

Preface to the First Edition

Most environmental concerns cross boundaries. Borders that separate countries, ecosystems, or jurisdiction of regulatory agencies are not respected by problems such as air pollution, declining anadromous fisheries, forest diseases, or threats to biodiversity. To address these problems, environmental planners and decision-makers must consider how geographically related systems are linked to form larger systems. Issues that may appear to be local will often require solutions at the landscape and region scale—working with the larger pattern, understanding how it works, and designing in harmony with it.

Following this reasoning, my task was to develop a geographical, ecologically based system, which would classify the natural ecoregions of the Earth and plot their distribution. The project built on work published by John Crowley in 1967 (Crowley 1967). In 1989, I published a terrestrial ecoregions map at a scale of 1:30,000,000 (Bailey 1989). A simplified, reduced-scale version appears in my book *Ecosystem Geography* (Bailey 1996) and the 19th edition of *Goode's World Atlas*. The influence of such ecoregion mapping efforts on research and planning efforts has been considerable. For example, the National Science Foundation's Long-Term Ecological Research program and The Nature Conservancy's ecoregional planning programs are taking place within the framework of this map. The Sierra Club announced in 1994 (*Sierra* March–April 1994) a "critical ecoregions" program designed to protect and restore 21 regional ecosystems in the United States and Canada. In another effort, the United States Forest Service has adopted ecoregion units as part of an ecologically based mapping system to support ecosystem management and assessment.

Understanding the continental systems requires a grasp of the ocean systems that exert enormous influence on terrestrial climatic patterns. This book is unique in the extended treatment (see Chaps. 2 and 3, and Plate 1) of oceanic ecoregions.

Ecoregions is intended to provide detailed descriptions, illustrations, and examples that will assist the user of the ecoregion maps in interpreting them. It amplifies the necessarily brief descriptions of the ecoregion units which appear in the legends to the maps. However, description without reference to genesis or origin soon becomes dull and tiresome—terms which unfortunately characterize much of the ecological literature on regions. A major objective of this book, therefore, is to suggest explanations of the mechanisms which act to produce the world pattern of ecoregion distribution, and to consider some of the implications for land use. The global extent of this book and its maps dictates that its ecoregion classification scheme be kept simple as possible, recognizing only principal ecoregion types. Where regional studies require additional detail, numerous additional subdivisions as needed can be added within an ecoregion type. For an example of detailed regional studies of ecoregions which follow this principle of creating subdivisions within the recognized world types, see "Ecoregions of the United States" (Bailey 1995).

Ecoregions was written for several audiences. In addition to the environmental planners and decision-makers mentioned at the beginning of this Preface, I hope that the increasing visibility of scale- and system-based science in ecological and environmental research will bring this work to the attention of workers in those fields. In addition, students and instructors should find the ecoregion approach useful in courses ranging from environmental planning to biogeography and ecosystem or landscape ecology.

I am indebted to Preston E. James, Günter Dietrich, John J. Hidore, Arthur N. Strahler and Alan H. Strahler, Heinrich Walter, and J. Schultz, for their own wonderful and insightful books.

I would like to acknowledge John M. Crowley, who began the work of global ecosystem regionalization. Recognition should also go to Chris Risbrudt, Director of Ecosystem Management Coordination in the Washington Office of the U.S. Forest Service, for his support. My thanks to Lev and Linda Ropes for helping me explain and illustrate the ideas in this book. The maps were made by Jon Havens. Susan Strawn made some of the drawings. As always, it has been a pleasure to work with Rob Garber at Springer-Verlag.

Fort Collins, CO   Robert G. Bailey
April 1998

Contents

The Colorado Plateau, Utah-Arizona, illustrates how a temperate steppe ecoregion is subdivided by the arrangement of surface features (which modify the regional climate) into sand dune, rock, and shrubland sites.

Introduction

Environmental problems are best addressed in the context of geographic areas defined by natural features rather than by political or administrative boundaries. For example, the state of Colorado in the western United States is neatly and abruptly divided into two areas with dramatically different ecological, climatological, and land-use characteristics: the eastern plains and western mountains (Fig. 1.1). Furthermore, both areas extend beyond Colorado's borders.

More and more we recognize that the natural resources of an area do not exist in isolation. Instead, they interact so that the use of one affects another (Fig. 1.2). This understanding has led entities responsible for the management of public—and increasingly, private land—to delineate and manage **ecosystems**,[1] rather than individual species or individual resources such as timber or range.

This approach recognizes that the Earth operates as a set of interrelated systems within which all components are linked. A change in one component causes a change within another with corresponding geographic distributions, as when certain vegetation and soil types occur together with certain types of climate. The regions occupied by tropical rainforest, for example, are found in tropical wet climates, and the underlying soils in the rainforest tend to be latosols (Oxisols)[2] (Fig. 1.3).

Ecosystems occur on many geographic scales. We can recognize ecosystems of different size, from oceans to frog ponds, and vast deserts to pockets of soil. The smaller systems are embedded, or nested, within larger systems. The larger systems are the environments of those within, controlling their behavior. By understanding the large forces that create macroscale ecosystems, we can predict how management practices will affect smaller component systems. For example, on a macroscale, the continents are embedded in the ocean systems that control them, through their influence on climatic patterns. A hierarchical ordering of the scales of ecosystems, from macro to micro, is presented in detail in my related book, *Ecosystem Geography: From Ecoregions to Sites* (Bailey 2009).

1.1 Concepts of Ecosystem Regions, or Ecoregions

Large portions of the Earth's surface over which the ecosystems have characteristics in common are called an ecosystem region, or **ecoregion**. Plates 1 and 2 (Maps, p. 165) show that areas

---

[1] Terms in bold are defined in the Glossary, p. 153.

[2] Great soil group according to the 1938 system of soil classification (U.S. Department of Agriculture 1938). The most nearly equivalent orders of the new soil taxonomy (USDA Soil Survey Staff 1975) are given in parentheses. Described in the Glossary, p. 153.

R.G. Bailey, *Ecoregions*, DOI 10.1007/978-1-4939-0524-9_1, © Springer Science+Media, LLC 2014

Fig. 1.1 Abrupt rise of the Colorado Front Range above the smooth surface of the North American Great Plains, looking north near Colorado Springs. Photograph by T.S. Lovering, U.S. Geological Survey

Fig. 1.2 Formation of gullies due to overgrazing, erosion, and increased runoff near Canberra, Australia. Grazing has eliminated the grass cover, reducing the retention of rainwater and facilitating the concentration of runoff. Photograph by Stanley A. Schumm, U.S. Geological Survey

with similar ecosystems are found in similar latitudinal and continental locations. Therefore, the distribution of ecoregions is not haphazard; they occur in predictable locations in different parts of the world and can be explained in terms of the processes producing them. For instance, temperate continental ecoregions in the Northern Hemisphere are always located in the interior of continents and on the leeward, or eastern, sides; thus the northeastern United States is in some ways similar to northern China, Korea, and northern Japan (Fig. 1.4 and Plate 2).

Because of this predictability, we can make assumptions about ecological features such as vegetation type that can be transferred across similar ecoregions of the same continent, or analogous ecoregions on different continents. Because data can be reliably extended to analogous sites within an ecoregion, we may greatly reduce data sampling and monitoring.

1.2   Need for a Comparative System of Generic Regions

Some schemes of classifying ecosystems have been based on the intuitive recognition of homogeneous-appearing regions, without considering the controlling forces that differentiate them. Using such methods each ecoregion is considered unique, unrelated to other regions. These are nothing more than "place name regions" such as the Great Plains of North America or the high Altiplano of Bolivia, instead of being based on criteria that define what type of region each is.

As a result of this, analogous regions in different continents or oceans may not be defined in the same way. Such inconsistency makes it difficult to exchange environmental information. Regions defined without specifying the factors upon which they were based are difficult for others to scrutinize or confirm. The results are therefore difficult to communicate convincingly. In this book I use a more explicit approach where regions are studied on the basis of comparable likenesses and differences. Such explicit methods require us to consider the physical factors that underlie ecosystem differentiation.

Understanding the processes involved in ecosystem (ecoregion) differentiation provides a basis for selecting significant criteria: those which are responsible for creating the range of ecoregion types found on the Earth. The purpose of this book is to describe and explain the character and arrangement over the Earth of the major ecosystem types, and the causes behind those patterns.

The face of the Earth could yield a nearly infinite variety of regional ecosystem types,

Tropical wet climates

Tropical rainforests

Latosolic soils

Fig 1.3 Spatial correspondence in the tropics between broad categories of climate, vegetation, and soils. Climate from Trewartha; vegetation after Eyre, Küchler, and others; soils based on numerous sources including Soil Conservation Service. From *Physical Elements of Geography,* 5th ed., by Glenn T. Trewartha, Arthur H. Robinson, and Edwin H. Hammond, Frontispiece, Plate 5, Plate 6. Copyright © 1967 by McGraw-Hill Inc. Reproduced by permission of McGraw-Hill, Inc.

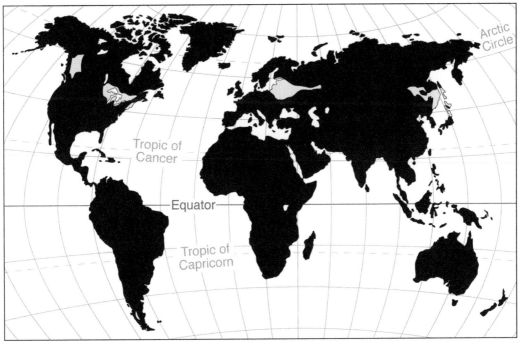

Warm continental zones

Fig. 1.4 Global pattern of the warm continental ecoregions

each defined by application of different criteria. But not all homogeneous areas are of equal significance. We are seeking meaningful types by the identification of correspondence between spatial patterns. That is, the manner in which the variations within one ecosystem component correspond to the variations within another component, particularly in ways that affect process. For example, all steep slopes have shallow soils and are susceptible to erosion.

1.3   The Process of Defining

The fundamental question facing all ecological mappers is: How are the boundaries of systems to be determined? The first, large-scale divisions of the Earth's surface are, quite obviously, the land masses and the water areas, where ecological processes take place in quite a different manner. But how do we determine the distribution of ecosystems within each?

Different methods have been used to identify units where ecosystem components (i.e., climate, vegetation, soil) are integrated in a similar way, thereby classifying land as ecosystems. The problem is boundaries on different component maps rarely correspond to each other. Further, overlay mapping techniques and cluster analysis address neither the causes that generate different ecosystem units nor why those units should be distinguished.

Establishing a hierarchy of ecosystem boundaries should be based on understanding of the **formative processes** that operate to differentiate the landscape into ecosystems at various scales (cf. Bailey 1985, 1987; Klijn and Udo de Haes 1994; Godron 1994). The units derived from such an approach are termed "genetic" in that they follow the cycle of landscape evolution (Fig. 1.5) and are predicated upon an understanding of the causal processes that control the pattern of ecosystems. *Understanding spatial relationships between causal mechanisms and resultant patterns is the key to understanding how ecosystems respond to management.*

The genetic approach is based on the idea of defining ecosystems using a deductive "top

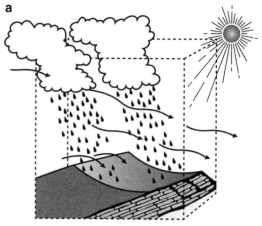

Climatic influences

Development of runoff system and landforms

Formation of soil, vegetation and habitat

Fig. 1.5 Schematic diagram illustrating the cycle of landscape evolution beginning with (**a**) climate, (**b**) runoff and landforms, and leading to (**c**) soil, vegetation, and habitat formation. From Marsh (2005), p. 58. Used with permission

down" approach. This is because subsystems can be understood only within the context of the whole; a mapping of ecosystems begins with the largest units and successively subdivides them.

The notion that process knowledge should define landscapes has a long history in geography and geology. William Morris Davis (1899) argued for a *genetic* classification of landforms, and both Herbertson's (1905) natural regions of the world and Fenneman's (1928) physiographic divisions of the United States are based on

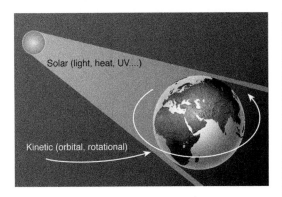

Fig. 1.6 Fundamental sources of energy that control conditions on Earth. Base from Mountain High Maps. Copyright © 1995 by Digital Wisdom, Inc.

Fig. 1.7 Petrified forest that now lies in a desert zone in Arizona. Photograph by D.B. Sterrett, U.S. Geological Survey

rational observation of underlying processes. Likewise, both Dryer (1919) and Sauer (1925) called for a genetic approach to landscape classification. The need to classify landscapes based on a reasoning of the causes of the phenomena as opposed to **empirical** classifications that are only descriptive and do not explain why regions should be distinguished is emphasized by Mackin (1963). Building on this work, I have developed, and continue to refine, a system of mapping ecosystems using the genetic approach (Bailey 1985, 1987, 1988, 1996, revised 2009), which spawned from concepts advanced by Crowley (1967). Delineating units involves identifying the factors thought to differentiate ecosystems at varying scales and drawing boundaries where they change significantly.

1.4 The Role of Climate

Energy is the prime driving force and controller of conditions on Earth. The three main sources of energy are (1) solar radiation, providing heat and light, (2) the kinetic energy of the rotation and orbit of the Earth, and (3) internal forces of both heat and kinetic energy. Figure 1.6 shows the first two sources. Internal forces will be discussed in Chap. 4.

As a result of the way, the Earth revolves about the sun and rotates on its axis, the low latitudes or tropics receive more solar radiation than do middle and higher latitudes.[3] Only about 40 % as much solar energy is received above the poles as above the equator. To balance this energy there is a large-scale transfer of heat poleward, which is accomplished through atmospheric and oceanic circulation. The frictional effects of the rotating Earth's surface on air flow cause these circulations to be relatively complex. Nevertheless, the solar energy and atmospheric and oceanic circulations are distributed over the Earth in an organized fashion. These controls, in turn, produce recognizable world patterns of temperature and precipitation, the two most important climatic elements.

Climate, as a source of energy and water, acts as the primary control for ecosystem distribution. As climate changes, so do ecosystems, as a petrified forest lying in a desert attests (Fig. 1.7).

1.5 Macroclimate

If the Earth had a homogeneous surface (either land or water), there would be circumferential zones of climate resulting from variations in the

---

[3] Zones of latitude may be described as follows: from the equator to 30° are *low latitudes*; from 30° to 60° are the *middle latitudes*; from 60° to the poles are the *high latitudes*.

Fig. 1.8 Latitudinal climatic zones that would result if the Earth were simply a granite sphere with an atmosphere

amount of solar radiation that reaches different latitudes (Fig. 1.8). These large climatic zones, or **macroclimates**, would be arranged simply in latitudinal bands, or east–west belts. They would owe their differentiation to the varied effects of the sun, instead of the character of the surface.

However, the Earth's surface is rather heterogeneous, divided first into large masses of land (the continents), water (the oceans), and ice (the polar regions). Each modifies the otherwise simple macroclimates. We first consider the oceans in the next chapter.

1.6   Advantages of the Ecoclimatic Approach

In contrast to empirical systems, a major emphasis in an **ecoclimatic** approach is on causal mechanisms that produce the patterns of ecosystem distribution. Understanding the mechanisms allows for (1) a comparative system of regions to be recognized, and (2) a certain degree of predictability and thus extrapolation of information (e.g., outcome of land use) from one geographic area to another. These two outcomes exemplify the benefits of establishing a hierarchy of ecosystem boundaries based on understanding formative processes.

This approach, which the U.S. Forest Service has adopted, is based on understanding the role of climate in ecosystem differentiation. Delineating

units involves analyzing controlling factors that operate to differentiate ecoclimatic units at different scales, and then using significant changes in controls as boundaries.

Advantages of this approach include:

- Recognizing ecoregions regardless of land use or anthropogenic disturbance
- Mapping the extent and character of ecoregion changes to display the impact of climate change on global ecoregion geography
- Using climate-driven ecoregion-based **fire regimes** to assess, restore, or alter **carbon sequestration**
- Extending sample information to similar ecosystems within the same ecoregion by recognizing the patterns of finer-scale ecosystems caused by landforms disrupting macroclimate

Many geographers and ecologists now understand the influence climate exerts on geographic areas and their ecosystems; see the bibliography section for recommended readings in this important area.

References

Bailey RG (1985) The factor of scale in ecosystem mapping. Environ Manage 9:271–276

Bailey RG (1987) Suggested hierarchy of criteria for multi-scale ecosystem mapping. Landsc Urban Plan 14:313–319

Bailey RG (1988) Ecogeographic analysis: a guide to the ecological division of land for resource management. Miscellaneous publication no. 1465. USDA Forest Service, Washington, DC, 16 pp

Bailey RG (1996) Ecosystem geography. Springer, New York, 204 pp

Bailey RG (2009) Ecosystem geography: from ecoregions to sites, 2nd edn. Springer, New York, 251 pp

Crowley JM (1967) Biogeography. Can Geogr 11:312–326

Davis WM (1899) The geographical cycle. Geogr J 14:481–504

Dryer CR (1919) Genetic geography. Ann Assoc Am Geogr 10:3–16

Fenneman NM (1928) Physiographic divisions of the United States. Ann Assoc Am Geogr 18:261–353

Godron M (1994) The natural hierarchy of ecological systems. In: Klijn F (ed) Ecosystem classification for environmental management. Kluwer Academic, Netherlands, pp 69–83

Herbertson AJ (1905) The major natural regions: an essay in systematic geography. Geogr J 25:300–312

Klijn F, Udo de Haes HA (1994) A hierarchical approach to ecosystems and its applications for ecological land classification. Landsc Ecol 9:89–109

Mackin JH (1963) Rational and empirical methods of investigation in geology. In: Albritton LC (ed) The fabric of geology. Addison-Wesley, Reading, MA, pp 135–163

Marsh WM (2005) Landscape planning: environmental applications, 4th edn. Wiley, New York, 458 pp

Sauer CO (1925) The morphology of landscape. Univ Calif Publ Geogr 2:19–53

Trewartha GT, Robinson AH, Hammond EH (1967) Physical elements of geography, 5th edn. McGraw-Hill, New York, 527 pp

U.S. Department of Agriculture (1938) Soils and men, 1938 yearbook of agriculture. U.S. Government Printing Office, Washington, DC, 1232 pp

U.S. Department of Agriculture, Soil Survey Staff (1975) Soil taxonomy: a basic system for making and interpreting soil surveys. Agriculture handbook 436. U.S. Department of Agriculture, Washington, DC, 754 pp

Oceanic Types and Their Controls

2

Oceans occupy some 70 % of the Earth's surface and extend from the North Pole to the shores of Antarctica. There are great differences in the character of the oceans, and these differences are of fundamental importance, both in the geography of the oceans themselves, and to the climatic patterns of the whole Earth. The surface of the ocean is differentiated into regions, or zones, with different hydrologic properties resulting from unevenness in the action of solar radiation and other phenomena of the climate of the atmosphere.

Ocean hydrology and atmospheric climate are both dynamic and closely interrelated. Ocean hydrology may be defined as the seasonal variation in temperature and salinity of the water. It controls the distribution of oceanic life. We use ocean hydrology to differentiate regional-scale ecosystem units and to indicate the extent of each unit. With continental ecosystems, atmosphere is primarily responsible for ecoregions but in the ocean, the physical properties of the water determine ecoregions, not the atmosphere.

2.1 Factors Controlling Ocean Hydrology

In order to establish the hydrographic zones of the world ocean, we must determine the boundaries of the zones. Our approach to this task is to analyze those factors that control the distribution of zones, and to use significant changes in those controls as the boundary criteria. This distribution in the character of the oceans is related to the following controlling factors:

2.1.1 Latitude

Heating depends predominantly on latitude. If the Earth were covered completely with water, thereby eliminating the deflection of currents by land masses, there would be circumferential zones of equal ocean temperature. The actual distribution forms a more complicated pattern (Fig. 2.1). Water temperatures range from well over 27 °C in the equatorial region, to below 0 °C at the poles. Areas of higher temperature promote higher evaporation and therefore higher salinity.

At low latitudes throughout the year and in middle latitudes in the summer, a warm surface layer develops, which may be as much as 500 m thick. Below the warm layer, separated by a thermocline, is a layer of cold water extending to the ocean floor. In Arctic and Antarctic regions, the warm surface water layer disappears and is replaced by cold water lying at the surface. The boundary between water types, called the **oceanic polar front**, is nearly always associated with a convergence of surface currents, discussed below.

2.1.2 Major Wind Systems

Prevailing surface winds set surface currents in motion and modify the simple latitudinal

R.G. Bailey, *Ecoregions*, DOI 10.1007/978-1-4939-0524-9_2, © Springer Science+Media, LLC 2014

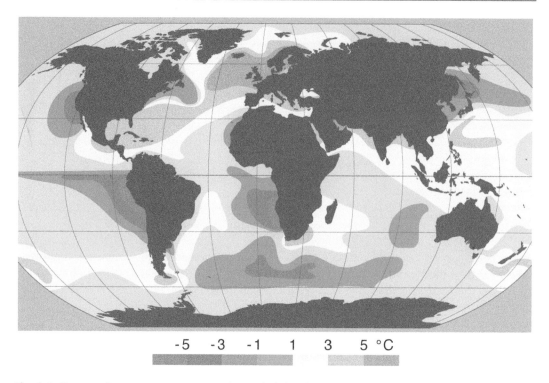

-5   -3   -1   1   3   5 °C

Fig. 2.1 Ocean surface temperatures expressed as a deviation from what they would be on a hypothetical globe covered only by water. From Dietrich (1963), Chart 3

arrangement of thermal zones. The most important of these winds, centered over the ocean basins at about 30° in both hemispheres, are known as **oceanic whirls** (Fig. 2.2). They circulate around the **subtropical high-pressure cells**, created by subsiding air masses (see Fig. 4.4, p. 30). Lying between the subtropical belts and the equator are easterly winds, known as **trade winds**. Near the equator the trade winds of both hemispheres converge to form a low-pressure trough, often referred to as the **intertropical convergence zone**, or ITC. On the poleward side of the subtropical highs is a belt of westerly winds, or **westerlies**. Generally, wind-driven ocean currents move warm water toward the poles and cold water toward the tropics, in a clockwise direction in the Northern Hemisphere and a counterclockwise direction in the Southern Hemisphere (Fig. 2.3).

The effects of these movements may be seen in Fig. 2.1. The average sea temperature on the coast of southern Japan, washed by the warm Kuroshio Current, is nearly 8° warmer than that in southern California, in the same latitude, but bathed by a cool current reinforced with **upwelling**. These are areas where ocean currents tend to swing away from the continental margins and cold water wells up from underneath.

The Earth's rotation, tidal swelling, and density differences further contribute to the dynamics of the oceans, both at the surface and well below. Distribution and diversity of ocean life are subsequently affected. Abundance is generally higher in areas of cooler temperatures and areas of upwelling of nutrients (Fig. 2.4).

2.1.3   Precipitation and Evaporation

The amount of dissolved salts, called **salinity**, varies throughout the oceans. It is affected by the local rates of precipitation and evaporation. Heavy rainfall lowers the surface salinity by dilution; evaporation raises it by removing water. Average surface salinity worldwide is about 35 parts per thousand (Fig. 2.5).

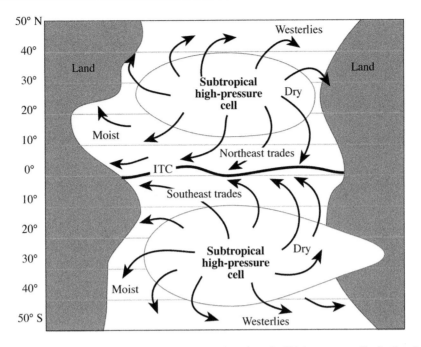

Fig. 2.2 Over the oceans, surface winds spiral outward from the subtropical high-pressure cells, feeding the trades and the westerlies. These drive a circular flow on the surface of the oceans. Adapted from Strahler (1965), p. 64; reproduced with permission

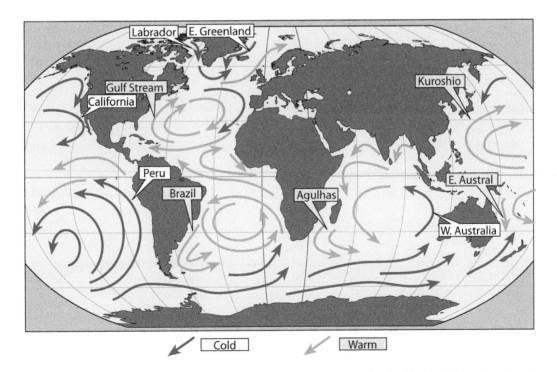

Fig. 2.3 The major oceanic surface currents are warm when flowing poleward and cold when flowing from the poles

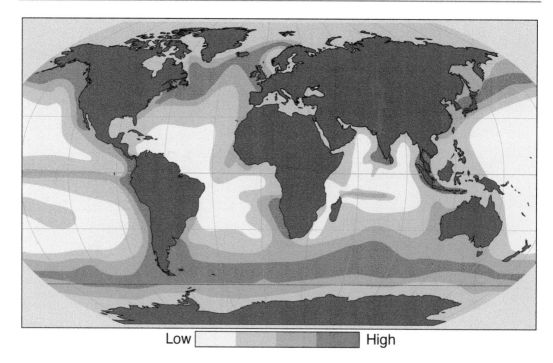

Fig. 2.4 The distribution of zones of productivity in the oceans is generally related to areas of cool water and upwelling. From Lieth (1964–1965)

The highest salinities in the open sea are found in the dry, hot tropics, where evaporation is great. Nearer the equator, salinities decrease because rainfall is heavier. In the cooler middle latitudes, salinities are relatively low because of decreased evaporation and increased precipitation. Surface salinities are generally low in the Arctic and Antarctic waters because of the effect of melting ice, which dilutes the sea water. In coastal waters and nearly enclosed seas, the salinity departs greatly from this pattern. In hot, dry seas of the Mediterranean and Middle East, the salinity is greater because the water is subject to strong evaporation and cannot mix readily with the open ocean. Near the mouths of large rivers, and in nearly enclosed seas fed by large rivers, such as the Baltic Sea (Fig. 2.5), dilution by fresh water reduces the salinity.

2.2 Types of Hydrologic Regions, or Oceanic Ecoregions

The interaction of the oceanic macroclimates and large-scale ocean currents determine the major hydrologic regions with differing physical and biological characteristics. These regions are defined as oceanic ecoregions. Their boundaries follow the subdivision of the oceans into hydrographic regions by Günter Dietrich (1963), although some modification has been made to identify regions with high salinity and the zonal arrangement of the regions. Dietrich's oceanographic classification and their ecoregion equivalents are summarized in Tables 2.1 and 2.2. Dietrich's system was used with similar results by Hayden et al. (1984) in their proposed regionalization of marine environments.

2.2.1 Dietrich's Oceanographic Classification

Dietrich's classification takes into account the circulation of the oceans, the temperature and salinity, and indirectly, the presence of major zones of upwelling. The motion of the surface is emphasized because of its influence on temperature. Salinity is generally higher in areas of higher temperature and therefore, higher

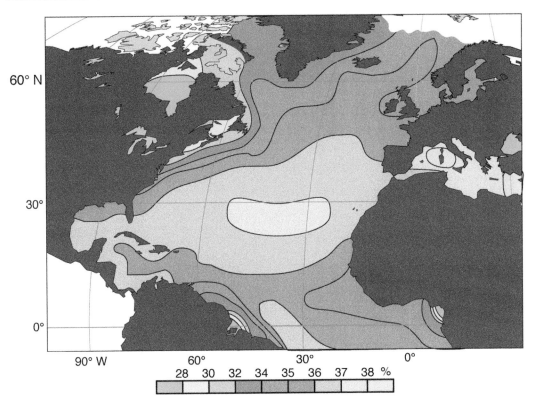

60° N

30°

0°

90° W   60°   30°   0°
                   28   30   32   34   35   36   37   38 %

Fig. 2.5 Surface salinities in the North Atlantic and adjacent waters are affected by the amounts of precipitation and evaporation. Modified from Dietrich (1963); in *Physical Elements of Geography* by Glenn T. Trewartha, Arthur H. Robinson, and Edwin H. Hammond, p. 392. Copyright © 1967 by McGraw-Hill Inc. Reproduced by permission of McGraw-Hill, Inc.

evaporation. Marine organisms are usually more abundant in cold waters, which makes the water appear more green.

There are seven main groups of hydrologic zones. Six are based on ocean currents, while the polar group is not. Four of these are subdivided into types based on current direction, latitude, and duration of pack ice. All together there are 12 regional divisions.

Dietrich (1963) describes them briefly. In the equatorial region is a belt with currents directed toward the east. These are the currents of the equatorial region, or the A group. On the low-latitude margins of the equatorial region are the currents of the trade wind regions, or the P group, where persistent currents move in a westerly direction. It is subdivided into three types, with a poleward current (Pp), a westerly current (Pw), or an equatorward component ($Pä$). Poleward from the trade winds is the horse wind region, or R group, where weak currents of variable direction exist. Still farther poleward is the region of west wind drift, or W group, where variable easterly currents prevail. Here, two subdivisions are recognized: poleward of the oceanic front (Wp) and equatorward of polar front ($Wä$). In the very high latitudes are the polar regions, or the B group, which are covered by ice. The duration of ice determines the subdivision: outer polar ($Bä$) in winter and spring which is covered with pack ice, and inner polar (Bi) which is covered with ice the entire year. The monsoon regions, group M, are subdivided into poleward monsoon (Mp) and a tropical monsoon (Mt). A final subdivision is the jet stream region, or F group, where strong, narrow currents exist as a result of discharge from trade wind regions.

Table 2.1 Natural regions of the oceans[a]

Dietrich group and types	Ecoregion equivalents
B **Boreal regions**	**Polar domain** (500)
Inner boreal (*Bi*)	Inner polar division (510)
Outer boreal (*Bä*)	Outer Polar division (520)
W **Westerly drift ocean regions**	**Temperate domain** (600)
Poleward of oceanic polar front (*Wp*)	Poleward westerlies division (610)
Equatorward of polar front (*Wä*)	Equatorward westerlies division (620)
R **Horse latitude ocean regions**	Subtropical division (630)
	High salinity subtropical division (640)[b]
F **Jet current regions**	Jet stream division (650)
M **Monsoon ocean regions**	
Poleward monsoons (*Mp*)	Poleward monsoon division (660)
	Tropical domain (700)
Tropical monsoon (*Mt*)	Tropical monsoon division (710)
	High salinity tropical monsoon division (720)[b]
P **Trade wind regions**	
With poleward current (*Pp*)	Poleward trades division (730)
With westerly current (*Pw*)	Trade winds division (740)
With equatorward component (*Pä*)	Equatorward trades division (750)
A **Equatorial ocean regions**	Equatorial countercurrent division (760)

[a]From the Dietrich system (1963)
[b]Not recognized by Dietrich; from Elliott (1954)

Table 2.2 Definitions and boundaries of the Dietrich system

P	(*Passat* in German), persistent westerly setting currents
Pä	With 30° equatorward component
Pw	Predominantly westerly set
Pp	With 30° poleward current
A	(*Aquator* in German), regions of currents directed, at times or all year to the east
M	(*Monsun* in German), regions of regular current reversal in spring and autumn
Mt	Low-latitude monsoon areas of little temperature variation
Mp	Mid to high (poleward) latitude equivalents of large temperature variations
R	(*Ross* in German), at times or all year marked by weak or variable currents
F	(*Freistrahlregionen* in German), all-year, geostropically controlled narrow current belts of mid-latitude westerly margins of oceans
W	(*Westwind* in German), marked by somewhat variable but dominantly east-setting currents all year
Wä	Equatorward of oceanic polar front (convergence)
Wp	Poleward of oceanic polar front (convergence)
B	At times or throughout the year, ice covered, in Arctic and Antarctic seas
Bi	Entire year covered with ice
Bä	Winter and spring covered with ice

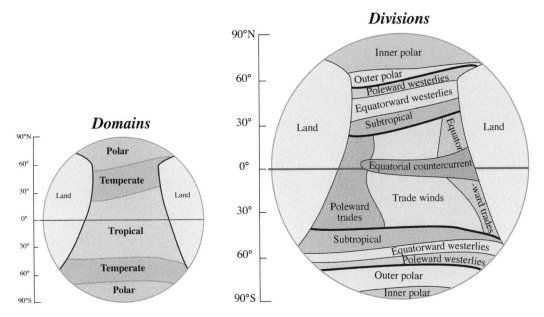

Fig. 2.6 Ecoregion domains and divisions in a hypothetical ocean basin. Compare with ocean map, Plate 1

2.3   Distribution of the Oceanic Regions

I have combined and rearranged the 12 regional divisions to identify oceanic regions. The resulting ecoregion **divisions** are shown diagrammatically in Fig. 2.6 as they might appear in a hypothetical ocean basin. On the diagram 12 kinds of regional divisions are recognized. They range from the inner polar division at high latitudes to the equatorial countercurrent division at low latitudes. Two of them are further subdivided (not shown on diagram) on the basis of high salinity following work by Elliott (1954).

I further simplified this classification of oceanic regions, using some of Schott's ideas (1936, as reported by Joerg 1935), by grouping the divisions in larger regions, called **domains** (Fig. 2.6). This recognizes the fact that the oceanic regions are arranged in latitudinal belts which reflect the major climatic zones. In both hemispheres there are three contrasting types of water, differing in temperature, salinity, life forms, and color. They are separated along the major lines of convergence where surface currents meet. In the high latitudes are the *polar waters*, characterized generally as low in temperature, low in salinity, rich in plankton, and greenish in color. These currents, also, frequently carry drift ice and icebergs. In the low latitudes are the *tropical waters*, generally high in temperature, high in salinity, low in organic forms, and blue in color. In between lie the so-called *mixed waters* of the temperate, middle latitudes.

As in the case of all phenomena on the Earth's surface that are directly or indirectly related to the climate, there is a tendency toward regularity in the pattern of arrangement of the hydrologic zones. However, the distribution of these various kinds of ocean water and the convergences that separate them is not strictly latitudinal. The east coasts of the continents are bathed by poleward-moving currents of tropical water. The west coasts at the same latitudes are bathed by colder waters moving from the poles. Because of the configuration of the ocean basins, the warm, east-coast currents of the Atlantic and the Pacific Oceans are best developed in the Northern Hemisphere, and the cold, west-coast currents in the Southern Hemisphere. As mentioned above,

Table 2.3 Approximate area and proportionate extent of oceanic ecoregions

	km^2	%
500 Polar domain	50,601,000	14.13
510 Inner polar division	20,305,000	5.67
520 Outer polar division	30,296,000	8.46
600 Temperate domain	136,620,000	38.15
610 Poleward westerlies division	17,977,000	5.02
620 Equatorward westerlies division	56,868,000	15.88
630 Subtropical division	52,929,000	14.78
640 High salinity subtropical division	1,217,000	0.34
650 Jet stream division	4,583,000	1.28
660 Poleward monsoon division	3,043,000	0.85
700 Tropical domain	118,714,000	33.15
710 Tropical monsoon division	9,239,000	2.58
720 High salinity tropical monsoon division	2,685,000	0.75
730 Poleward trades division	18,299,000	5.11
740 Trade winds division	54,325,000	15.17
750 Equatorward trades division	14,790,000	4.13
760 Equatorial countercurrent division	19,373,000	5.41
Shelf	52,177,000	14.57

where currents tend to swing offshore, cold water wells up from below.

In my system, there are four domains, and 14 divisions, plus the continental shelf area where the shallow waters (<200 m), are interpreted as shallow variations of the ecoregion involved. Plate 1 (Maps, p. 165) shows the distribution of the oceanic ecoregion domains and divisions. Table 2.3 presents a rough estimate of the area contained in each region and its percentage of the total ocean area.

Each domain/division occurs in several different parts of the world that are broadly similar with respect to physical and biological characteristics. For example, the equatorial trades division (750) occurs only on the west sides of large continents and is set off from the adjacent low-latitude oceanic waters by lower temperatures, somewhat wider variations of temperatures, and lower salinity.

We will now treat the characteristics, extent, and subdivisions of each of the regions distinguished in the next chapter. Their characterization provides further information on the principles which has been applied in drawing the boundaries in Plate 1. The descriptions of the characteristics are drawn and summarized from several sources; the most important is

Dietrich (1963), with supplementary information by Elliott (1954), and James (1936).

References

Dietrich G (1963) General oceanography: an introduction. Wiley, New York, 588 pp

Elliott FE (1954) The geographic study of the oceans. In: James PE, Jones CF (eds) American geography: inventory & prospect. Association of American Geographers by Syracuse University Press, Syracuse, NY, pp 410–426

Hayden BP, Ray GC, Dolan R (1984) Classification of coastal and marine environments. Environ Manage 11:199–207

James PE (1936) The geography of the oceans: a review of the work of Gerhard Schott. Geogr Rev 26:664–669

Joerg WLG (1935) The natural regions of the world oceans according to Schott. In: Transactions American Geophysical Union, Sixteenth Annual Meeting, Part I; 25–26 April 1935; Washington, DC, pp 239–245

Lieth H (1964–1965) A map of plant productivity of the world. In: Geographisches Taschenbuch, Wiesbaden, pp 72–80.

Schott G (1936) Die aufteilung der drei ozeane in natürliche regionen. Petermann's Mitt 82 (165–170):218–222

Strahler AN (1965) Introduction to physical geography. Wiley, New York, 455 pp

Trewartha GT, Robinson AH, Hammond EH (1967) Physical elements of geography, 5th edn. McGraw-Hill, New York, 527 pp

Ecoregions of the Oceans

3.1   500 Polar Domain

At times, or during the winter, ice of the Arctic or Antarctic Oceans covers these regions. They are characterized, in general, by ocean water that is greenish, low in temperature, low in salt content, and rich in small, sometimes microscopic plant and animal organisms, known as **plankton**. The duration of ice provides a basis for division into (a) an *inner polar zone* (division) covered by ice for the entire year, and (b) an *outer polar zone* (division) where, with a 50 % probability, ice is encountered during winter and spring. Figure 3.1 shows the global locations of the polar ecoregions of the oceans.

3.1.1   510 Inner Polar Division

In the Arctic region, the inner polar region includes the deep Arctic basin and the northern passages of the Canadian Archipelago; in the Antarctic region it includes a narrow belt around the Antarctic ice shelf. This is the region that Dietrich (1963) designates as *Bi*. Although the region is always covered with ice, the ice itself is in continuous motion, clockwise around the poles. In the Northern Hemisphere this area is free of icebergs, because no glaciers touch this region. Hummocked pack ice forms impenetrable ridges with intermittent leads and openings here and there, due to severe ice pressure. In contrast, the central parts of the inner polar region are relatively flat, pack-ice fields.

The Arctic Ocean, which is surrounded by landmasses, is normally covered by pack ice throughout the year, although open leads are numerous in the summer. The relatively warm North Atlantic drift maintains an ice-free zone off the northern Coast of Norway. The situation is quite different in the Antarctic, where a vast, open ocean bounds the sea ice zone on the equatorward margin. Because the ice flows can drift freely north into warmer waters, the Antarctic ice pack does not spread far beyond 60°S latitude in the cold season.

Air temperatures are generally below freezing, and the annual variation of air temperature is large. Because of the low air temperatures and the resultant low capacity of the air to hold moisture, the annual amount of precipitation is very small, but exceeds evaporation.

3.1.2   520 Outer Polar Division

Adjacent to the inner polar region lies the outer polar region, which is regularly covered by pack ice during winter in both hemispheres. It is classified as *Bä* in the Dietrich system. Pack ice does not originate within this region. It represents ice carried into this region by oceanic currents from the inner polar region, and forms fields of drifting ice that will gradually melt. Oceanic currents considerably influence the position of the boundaries of this region. With the cold East

R.G. Bailey, *Ecoregions*, DOI 10.1007/978-1-4939-0524-9_3, © Springer Science+Media, LLC 2014

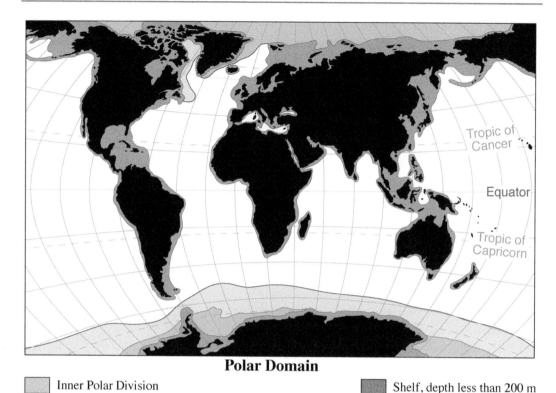

Polar Domain

Inner Polar Division   Shelf, depth less than 200 m
Outer Polar Division

Fig. 3.1 Divisions of the oceanic polar domain

Greenland Current and the Labrador Current, the polar pack ice is carried far to the south end of the Grand Banks (46°N). In contrast, the extension of the warm Gulf Stream system keeps ice-free during the entire year, the eastern side of the Norwegian Sea and the southern Barents Sea. This allowed convoys from America during World War II to reach the Russian (formerly Soviet) port of Murmansk.

In summer the hydrologic conditions of the *Bä* regions do not differ considerably from those of the *Wp* regions. This is especially true for the Southern Hemisphere where variable, predominantly easterly currents are common to both regions. The boundaries between the two regions, as well as the boundary between the two polar regions, favor the development plankton. In the feeding chain, plankton forms the basis of krill shrimp on which, the blue whales and finback whales feed. These relationships explain why, in summer, the two boundaries

become the main feeding grounds of the whales, and consequently the main whaling grounds.

3.2 600 Temperate Domain

This comprises the middle latitudes between the poleward limits of the tropics and the equatorward limits of pack ice in winter. Currents in this region correspond to wind movements around the subtropical, high-pressure cells of the atmosphere. These are the so-called mixed waters of the middle latitudes. The variable direction of the current determines the divisions: (a) a *poleward westerlies zone* (division) with cold water and sea ice that is poleward of the oceanic polar front; (b) an *equatorward westerlies zone* (division) that has cool water and is equatorward of the oceanic polar front; (c) a *subtropical zone* (division) with weak currents of variable directions; (d) a *high-salinity, subtropical zone*

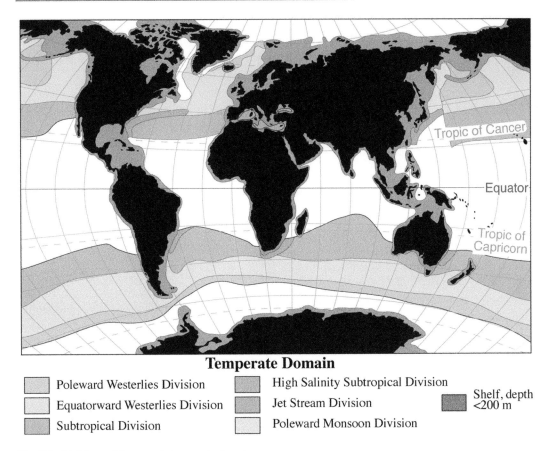

Temperate Domain

<table>
<tr><td>☐ Poleward Westerlies Division</td><td>☐ High Salinity Subtropical Division</td><td rowspan="2">■ Shelf, depth <200 m</td></tr>
<tr><td>☐ Equatorward Westerlies Division</td><td>☐ Jet Stream Division</td></tr>
<tr><td>■ Subtropical Division</td><td>☐ Poleward Monsoon Division</td><td></td></tr>
</table>

Fig. 3.2 Divisions of the oceanic temperate domain

(division) characterized by excess evaporation over precipitation; (e) a *jet-stream zone* (division) characterized by strong, narrow currents which exist during the entire year as the result of discharge from trade wind regions; and (f) a *poleward monsoon zone* (division) of high latitudes with the reversal of current (connected with large annual variations in surface temperature). A global map of the temperate ecoregions of the oceans is shown in Fig. 3.2.

3.2.1   610 Poleward Westerlies Division and 620 Equatorward Westerlies Division

These regions are characterized by variables, predominantly easterly currents during the entire year. The polar boundary is that zone which, during winter, is covered permanently or frequently with ice of the polar seas. In winter, east-traveling cyclones are strongly developed and produce storm zones. These include the "roaring forties" between 40° and 50° in the Southern Hemisphere as well as storm zones in the North Pacific and North Atlantic Oceans.

In the *W* regions in the Dietrich system, precipitation falls during all seasons and higher and more frequently in fall and winter. Precipitation exceeds evaporation, decreasing the surface salinity. The oceanic polar front lies within the *W* regions. In the Southern Hemisphere it coincides with the zones of strongest westerly winds. At the sea surface it separates the cold-water sphere from warm water. Along this front, the spreading of ice of the polar seas generally reaches its equatorial limit. This provides a basis to subdivide the *W* regions into a *Wp* region

Fig. 3.3 Elephant seals and penguins on the Crozet Islands, South Indian Ocean. Photograph by Douglas Mawson; from the American Geographical Society Library, University of Wisconsin-Milwaukee Libraries

(poleward of the polar front) and a *Wä* region (equatorward of the polar front).

Deep-reaching mixing processes are present in these regions. The associated high-nutrient concentrations provide an opportunity for plankton to develop abundantly. Plankton provides nutrition for great numbers of commercial fish, especially on the continental shelves. The shelves provide other sources of nutrition and also serve as spawning grounds. For this reason, the Grand Banks, the banks west of Greenland, around Iceland, and the Faroes as well as the shelf of Europe have become the main grounds of high-sea fishing. Oceanic islands in these regions, such as the Crozet Islands in the South Indian Ocean, have Antarctic birds and mammals which exist because of food chains that start in these seas (Fig. 3.3).

3.2.2   630 Subtropical Division

Between the region of the trade wind currents, *P* in the Dietrich system, and the region of the westerlies, lies a region of transition designated by the symbol, *R*. In this region, weak currents of variable direction exist. This region is associated with the subtropical high, which is associated with weak winds—called the *horse latitudes* at about 25°–30° north and south of the equator.

The horse latitudes are said to be so called from the throwing overboard of horses in transport from Europe to North America if the ship's passage were delayed by calms.

Trade winds and westerlies flow around the interior region of the horse latitudes. This causes an accumulation of light surface water in the center of the eddy. A deep, homogeneous, warm, top layer is established, which is also very saline because of the excess of evaporation over precipitation (see Fig. 2.5, p. 13). No other regions in the world ocean have higher temperature and salinity.

The *R* regions are also characterized by an extremely low nutrient content since planktons have consumed the nutrients. Therefore, the biomass assumes its smallest value in the surface waters, such as in the Sargasso Sea in the central *R* region of the North Atlantic Ocean. In contrast, the absolute maximum of organic production in the ocean has been found in the region of upwelling off Southwest Africa (see Fig. 2.4, p. 12).

As a consequence of very low plankton content, these regions are distinguished by extremely clear, transparent water, which shows a deep cobalt blue color. They are generally smooth seas because storms rarely touched them.

3.2.3   640 High-Salinity Subtropical Division

The high-salinity regions are not designated as a separate variety by Dietrich, which Elliott (1954) designates as type *H*. They have been incorporated into the system presented in this book. These regions are represented by the Mediterranean Sea, the Red Sea, and the Black Sea. Continental influences are extremely strong because the individual units are almost landlocked and water exchange with the oceans is very slow. Each unit shows a strong individuality, but nevertheless certain characteristics permit grouping them into one major type. These common characteristics are: almost landlocked position, extremely strong continental influence, wide range and variation of temperatures, wide range of salinities, wide variation of air temperature, great range of precipitation, and generally counterclockwise surface circulation. All are characterized by excess of evaporation over precipitation; making them highly saline (see Fig. 2.5, p. 13).

3.2.4   650 Jet Stream Division

Along the west sides of the oceans in low latitudes, the equatorial current turns poleward, forming a warm current paralleling the coast, indicated by the symbol F in the Dietrich system. These discharge currents are known as the Gulf Stream and Kuroshio in the Northern Hemisphere, and as the Brazil Current, East Austral Current, and Agulhas Current in the Southern Hemisphere. They bring higher than average temperatures along these coasts. The high velocity of these currents tends to cause a cross circulation. At the left side of the Gulf Stream, for example, the cross circulation brings water rich in nutrients from deeper layers to subsurface layers. This is associated with an abundance of plankton, causing the greenish-blue color of this adjacent water.

3.2.5   660 Poleward Monsoon Division

Dietrich classified the poleward monsoon waters as Mp. These regions occur mainly in east-Asiatic waters, where regular changes in the direction of the monsoon winds in spring and fall cause a reversal of the surface currents. The season of the winter monsoon lasts from November to March. During this time, under the influence of cooling, a strong atmospheric high develops over Asia. The air in the lowest layers flows out over the east-Asiatic marginal seas (i. e., the Sea of Japan, the Sea of Ochotsk, and the Bering Sea) from north to northwest. During the summer, from May to September, the flows are reversed on shore toward the continent.

The off-shore winds of the winter monsoon carry cold continental air over the seas, whereas the on-shore winds of the summer bring warm, oceanic air masses into these regions. Due to

these atmospheric influences, the sea surface temperatures have the greatest annual variations found in any ocean. In North Korean and Manchurian waters, variations of over 20 °C, sometimes even 25 °C are found.

In winter the water temperature drops to the freezing point. Even Vladivostok, Russia, on the latitude of Florence, Italy, does not remain ice free. In spring and summer, when the maritime air masses saturated with water vapor arrive monsoon-like over these cold oceanic areas, persistent sea fogs develop.

Comparative low salinities are caused by heavy runoff from the land, great excesses of precipitation over evaporation, and inflow of low-salinity, polar continental waters.

3.3   700 Tropical Domain

This is characterized by ocean water that is generally blue, high in temperature, high in salt content, and low in organic forms. It is divided into (a) *tropical monsoon* with regular reversal of the current system (connected with small annual variations in surface temperature); (b) *high salinity, tropical monsoon* with alternating currents; (c) *poleward trades* with a strong velocity directed toward the poles; (d) *trade winds* with current moving toward the west; (e) *equatorward trades* with a strong velocity directed toward the equator, and where currents tend to swing offshore and cold water well up from below; and (f) *equatorial countercurrents* where currents are directed at times or during the entire year toward the east. The global extent of the tropical ecoregions of the oceans is presented in Fig. 3.4.

3.3.1   710 Tropical Monsoon Division

Located over the North Indian Ocean and the waters around southeast Asia, is a region affected by monsoonal winds, which Dietrich designates as Mt. In winter, the winds blow from the north and northeast. In this season, the hydrologic conditions resemble those of a region of trade wind currents. The surface water flows toward

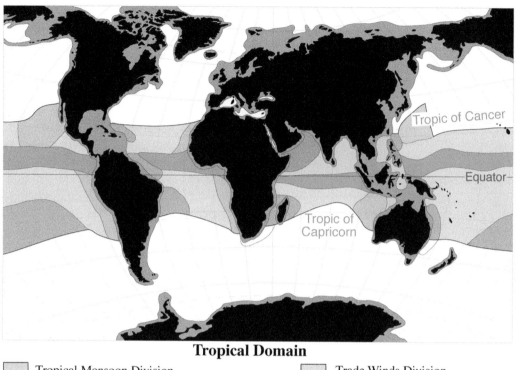

Tropical Domain

Tropical Monsoon Division
High Salinity Tropical
Monsoon Division
Poleward Trades Division

Trade Winds Division
Equatorward Trades Division
Equatorial Countercurrent Division
Shelf depth less than 200 m

Fig. 3.4 Divisions of the oceanic tropical domain

the west and becomes more saline along its way because the dry continental air does not produce precipitation, and evaporation is high.

Annual temperature variation is quite small; however, currents, winds, salinities, and precipitation show marked seasonal change because of the monsoon winds, which cause a reversal of the currents with the season.

3.3.2 720 High-Salinity Tropical Monsoon Division

This type is restricted to the Arabian Sea where dry winds bring about very low precipitation, strong evaporation, and resulting high salinities. In type 710, designated as type *E* by Elliott, salinities are considerably lower because of the monsoon rains and high runoff.

3.3.3 730 Poleward Trades Division

North and south of the equator are the currents of the trade wind belts, covering roughly the zones lying between 5° and 30°N and S. Its poleward boundaries follow approximately the mean stand of the longitudinal axes of the subtropical high-pressure cells (see Fig. 4.5, p. 31). Air moving equatorward from the cells is deflected by the Earth's rotation to turn westward, with currents running the same direction the entire year. Turning of the current (see Fig. 2.2, p. 11) is controlled partly by the turning of the trade winds and partly by the distribution of the continents. Those winds deviate the currents from the zonal direction in the eastern and western regions.

In the west of the division are regions of the trade wind currents with components directed toward the poles. These regions, which Dietrich designated at *Pp* are distinguished by anomalously high surface temperatures, which, together with the high evaporation, contribute to instability of the atmosphere and precipitation. If orographic rain also falls on the coast, the amounts are sufficient to develop lush vegetation on oceanic islands, such as the Polynesian Islands and the continental coast of Brazil, south of Bahia, as well as northeastern Australia.

3.3.4   740 Trade Winds Division

In these regions, the North and South Equatorial Currents, which Dietrich classified as *Pw*, move westward uniformly and persistently. The uniformity of current corresponds to uniformity of wind and weather. These regions have little precipitation and since evaporation is high, the sea surface is highly saline. Because the region is free of divergences and the annual variation of surface temperature is small, there is no vertical mixing to renew nutrients from deeper layers. Because the plankton population is, therefore, low, only a few higher marine organisms exist here.

High sea-surface temperatures—above 27 °C in these latitudes—are important in the development of tropical storms, which originate in this area. Warming of air at low levels creates instability and predisposes the area toward the formation of storms. Once formed, the storms move westward through the trade wind belt. They are by far the most violent storms on the Earth and are known as *typhoons* in East Asia, as *Mauritius hurricanes* in the South Indian Ocean, and as *hurricanes* in the West Indies.

3.3.5   750 Equatorward Trades Division

This type consists of currents known as the Canaries and Benguela Currents in the North and South Atlantic Oceans, as California and Peru (Humbolt) Current in the North and South Pacific Oceans, and as the West Australian Current in the South Indian Ocean. In the Dietrich system, areas of this type are designated as *Pä*. Since they flow from higher to lower latitudes, they carry water that is colder than average for the corresponding latitudes. These temperature differences become larger because the winds which blow parallel to the coast, deflect surface water seaward, thereby causing **upwelling**. This brings cold water, abundant in nutrients, to surface layers. It contributes to the extraordinary development of plankton. The abundance of plankton shows in the green coloring of the sea water, in contrast to the cobalt blue of the neighboring regions of the horse latitudes. The great abundance of plankton produces large amounts of fish in these areas, attracting endless numbers of sea birds. They are responsible for the formation of guano deposits at these coasts. When this upwelling is disturbed or replaced by a motion directed toward the pole (the so-called El Nino at the north Peruvian coast, which is not associated with upwelling but carries warm, nutrient depleted, equatorial water), then masses of fish, and subsequently birds, die.

In general, surface temperatures are lower than air temperatures, especially in the vicinity of coasts. Hence, the formation of lasting and frequent fogs, known under the name Garua at the Peruvian coast, is common. Low rainfall is the result of the stabilization of the air mass over the cold water surface. These areas have the lowest precipitation on the globe, including inland deserts. Oceanic islands in these regions, such as the Cape Verde Islands in the North Atlantic Ocean and the Galapagos Islands in the South Pacific Ocean, have desert-type climates. The coastal deserts of Namib in Southwest Africa and of Atacama in north Chile are located at approximately the same latitude adjacent to these cold currents.

Another consequence of the low surface temperature of these regions is the fact that no coral reefs are found in these areas. Their development requires not only clear water but also temperatures of at least 20 °C in the coldest month of the year (Fig. 3.5).

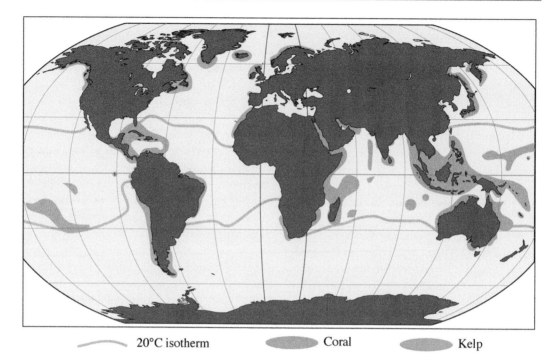

——— 20°C isotherm   ⬭ Coral   ⬭ Kelp

Fig. 3.5 Coral reefs are most abundant in the oceans between the 20 °C mean annual surface water temperature isotherm, whereas kelp is most abundant outside that zone. After K.H. Mann, seaweeds: their productivity and strategy for growth. *Science* 182, p. 976, 1973 and J.W.

Wells, Coral reefs, *Geological Society of America Memoir* 67 1, p. 630, 1957; in *Ecology of World Vegetation* by O.W. Archibold, figure 12.17, p. 405, Copyright © 1995 by Chapman & Hall. Reproduced with permission

3.3.6 760 Equatorial Countercurrent Division

The equatorial currents are separated by an equatorial countercurrent and are designated *A* by Dietrich. This condition is well developed in the three oceans. This region, lying roughly 5°S and 5°N lat., coincides with a trough of low pressure, the intertropical convergence zone (see also Chap. 4, p. 27 and Chap. 8, p. 81). The northeast and southeast trade winds come together toward this trough. It is a zone of variable winds and calms, or the **doldrums**. The zone contains large amounts of moisture, and cloudiness and frequent precipitation are common. The heavy precipitation lowers the salinity of the sea surface considerably.

Upwelling at the poleward side of the countercurrent results in ascending water masses rich in nutrients. As it reaches the surface, an abundant plankton population develops (see Fig. 2.4, p. 12). Like the areas of upwelling in the eastern region of the trade wind currents, the equatorial countercurrent zones are characterized by greenish color and an abundance of fish.

3.4 Shelf

All shallow-shelf areas with depths of 0–200 m are interpreted as shallow variations of the hydrologic zone concerned. A further ecoregional breakdown of coastal and shelf areas of the world is presented by Spalding et al. (2007). A controlling factor approach to classification (see Chap. 2) of estuaries in New Zealand is described by Hume et al. (2007). The classification is based on the principle that particular factors are responsible for environmental processes and patterns that are observed at various spatial scales. The classification differentiates estuaries at four levels of detail. The authors propose that the ocean ecoregions presented in this book, which defines regions of homogeneous climate and oceanic water masses at broad scales, provide an appropriate subdivision of estuaries at level 1 of the classification, a category based on the ocean ecoregion within which it is located.

References

Archibold OW (1995) Ecology of world vegetation. Chapman & Hall, London, 510 pp

Dietrich G (1963) General oceanography: an introduction. Wiley, New York, 588 pp

Elliott FE (1954) The geographic study of the oceans. In: James PE, Jones CF (eds) American geography: inventory & prospect. Association of American Geographers by Syracuse University Press, Syracuse, NY, pp 410–426

Hume TM, Snelder T, Weatherhead M, Liefting R (2007) A controlling factor approach to estuary classification. Ocean Coast Manage 50:905–929

Spalding MD, Fox HE, Allen GR, Davidson N (2007) Marine ecoregions of the world: a bioregionalization of coastal and shelf areas. Bioscience 57:573–583

Continental Types and Their Controls

4

The ecoregions on the Earth's land masses are arranged in predictable patterns and are causally related to macroclimate, i.e., the climate that lies just above the local modifying irregularities of landform and vegetation. These macroclimates are regularly arranged with reference to several controlling factors.

4.1 The Controls of Macroclimate

The controls of macroclimate may be grouped under three headings: latitude, continental position, and elevation.

4.1.1 Latitude

As mentioned previously, if the Earth had a homogeneous surface circumferential zones of macroclimate would result from the variation in solar radiation and the resultant atmospheric circulation. These zones would be divided along lines of latitude (see Fig. 1.8, p. 6).

Thermally Defined Zones The actual distribution of land and sea forms a more complicated thermal distribution (Fig. 4.1). The thermal limits for plant growth determine boundaries of these zones. For example, trees in Eurasia and America cannot grow beyond about 70° latitude.

We can delineate three major thermally defined zones: (1) a winterless climate of low latitude, (2) a temperate climate of mid latitudes with both summer and winter, and (3) a summerless climate of high latitude. In winterless climate, no month of the year has a mean monthly temperature lower than 18 °C. The 18 °C isotherm approximates the position of the boundary of the poleward limit of plants characteristic of the humid tropics, such as palms. In summerless climate, no month has a mean monthly temperature higher than 10 °C. The 10 °C isotherm closely coincides with the northernmost limit of tree growth, separating the regions of boreal forest (**tayga**[1]) from the treeless tundra.

If we also consider the annual and diurnal energy cycles, we can differentiate these thermal zones. The relative amplitudes of annual and diurnal energy cycles vary in each region (Fig. 4.2). Within the tropics, the diurnal range is greater than the annual range. Within temperate zones, the annual range exceeds the diurnal range, although the diurnal can be very large. Within the polar zones, the annual range is far greater than the diurnal range.

Moisture-Defined Zones Life-giving precipitation is generally higher in areas of higher temperature and, therefore, higher evaporation. It ranges from about 200 cm in the equatorial region to about 30 cm at the poles (Fig. 4.3).

---

[1] In this book tayga, often spelled "taiga," will be written with a "y" following the Russian.

R.G. Bailey, *Ecoregions*, DOI 10.1007/978-1-4939-0524-9_4, © Springer Science+Media, LLC 2014

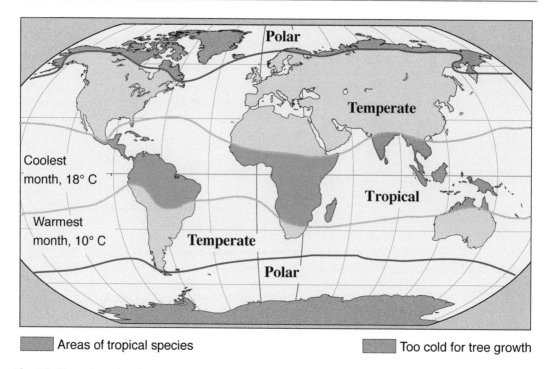

Coolest
month, 18° C

Warmest
month, 10° C

Areas of tropical species   **Too cold for tree growth**

Fig. 4.1 Zones determined by thermal limits for plant growth. From Strahler (1965), p. 103; redraw with permission

Precipitation and runoff also follow a zonal pattern, generally decreasing with latitude. Near the equator the trade winds converge toward the **intertropical convergence zone** (ITC). The trade winds moving toward the equator pick up moisture over the oceans and, when lifted in the ITC, yield abundant precipitation.

On top of this, differential heating at different latitudes causes transfer of heat from lower to higher latitudes, partly through circulation of the atmosphere. The result is a series of belts of ascending and subsiding air masses (Fig. 4.4). Subsiding air masses of the subtropical high-pressure cells, roughly centered on the tropics of Cancer and Capricorn, have adequate heat but a shortage of moisture. These zones are too dry for tree growth.

4.1.2   Continental Position

The interaction of land and sea modifies the situation. This division of the Earth's surface between land and sea, each with quite different thermal characteristics, results in distinctive differences between marine-influenced and inner-continental climates. For a hypothetical continent of uniform elevation, temperature distribution would look like that shown in Fig. 4.5.

In addition, the distribution of land and sea forms complicated but predictable, precipitation patterns—less precipitation over margins of continents bathed by cold water. The dry zone, controlled by the subtropical high-pressure cells, is shifted to the west side of the continents, adjacent to these cold currents (Fig. 4.6). These dry zones are too dry for trees and consist of deserts and grasslands. This creates a distribution of dry climatic zones that strongly affect ecosystem distribution.

Combined Latitude (Thermal) and Moisture Considerations By combining the thermally defined zones with the moisture zones, we can delineate four ecoclimatic zones or ecoregions: *humid tropical*, *humid temperate*, *polar*, and *dry*

Fig. 4.2 Variation in air temperature through the day and through the year in tropical, temperate, and polar zones. Stations are Singapore, Oxford, England, and McMurdo Sound, Antarctica, respectively. From Troll (1966), p. 15–16

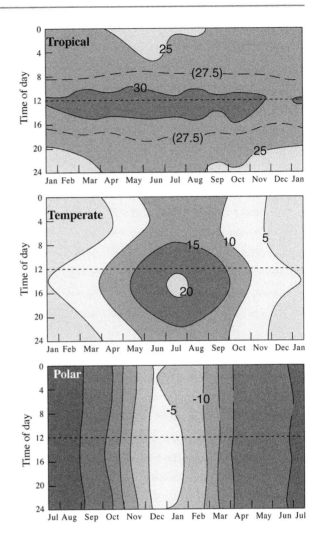

(Fig. 4.7). They are arranged in a regular, repeated pattern with reference to latitude and position on the continents.

Within each of these zones, one or several climatic gradients may affect the potential distribution of the dominant vegetation. Within the humid tropical zone, for example, we can distinguish rainforests that have year-round precipitation from savannas that receive seasonal precipitation. Thus, we can subdivide the humid tropical zone, based on moisture distribution (Fig. 4.8) into *climatic subzones*. We can subdivide the other zones similarly.

Locating the boundaries of broad-scale ecosystems requires taking into account visible and tangible expressions of climate such as vegetation. Generally, each climate is associated with a single **plant formation class** (such as savanna, see Fig. 8.5, p. 84, Table 4.1), and is characterized by a broad uniformity both in appearance and in composition of the dominant plant species. Usually a significant correspondence with soils occurs because climate also strongly dominates soil-forming processes. Of course, not all the area is taken up by the formation, because the nature of the topography will allow the differentiation into many sites. One ignores these local variations in mapping climatic regions (and therefore ecoregions).

The vegetation conditions indicated by plant formation name refer to undisturbed plant cover that is known to exist or is assumed to grow if

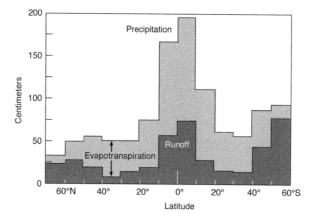

Fig. 4.3 Distribution of annual precipitation and runoff amounts averaged by latitudinal zones. The vertical difference between the two lines represents the loss through evapotranspiration. After L'vovich and Drozdov; from

Physical Elements of Geography 5th ed., by Glenn T. Trewartha, Arthur H. Robinson, and Edwin H. Hammond, p. 413. Copyright (c) 1967 by McGraw-Hill, Inc. Redrawn by permission of The McGraw-Hill Companies

human intervention should be removed. This is known as **potential natural vegetation** (Küchler 1964).

4.1.3 Modified by Elevation

The arrangement of the ecological zones is largely dependent on latitude. To further complicate matters, the Earth's internal energy causes irregular patterns of high mountains on the continents (Fig. 4.9). These modify and distort the simple climatic pattern that would develop on a flat continent. We can see the short-term and small-scale implications of this difference in local meteorological boundary effects (Fig. 4.10).

These mountains are arranged without conforming at all to the orderly latitudinal zones of climate. They cut irregularly across latitudinally oriented climatic zones. For example, we find mountains in the cold deserts of Antarctica as well as near the equator (Fig. 4.11). The regions of this type do not appear on the diagram showing the generalized global pattern of ecoregions (see Fig. 4.14, p. 39) because these features, along with the outlines of the continents, are unique for each land mass.

Mountains have a typical sequence of elevational belts, with different ecosystems at successive levels (Fig. 4.12). These differ

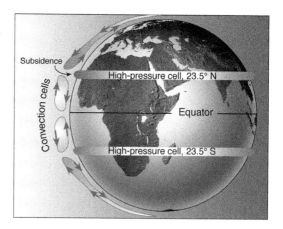

Fig. 4.4 The dry zones are controlled by the subtropical high-pressure cells that are caused by subsidence between the atmospheric convection cells. Base from Mountain High Maps. Copyright © 1995 by Digital Wisdom, Inc.

according to the zone in which the mountain is embedded. In other words, elevation produces a predictable variation of the lowland climate, especially in **climatic regime** (i.e., seasonality of temperature and precipitation). The coast ranges of California, for example, experience the same strong seasonal energy variations, and a seasonal moisture regime consisting of a dry summer and a rainy winter typical of their neighboring lowlands.

When a mountain occurs in two climatic zones, it produces different vertical zonation

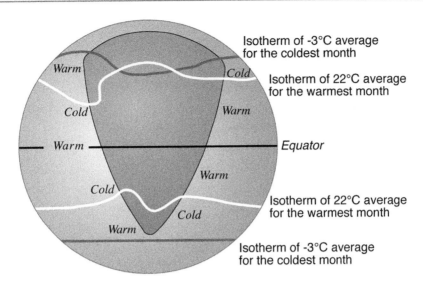

Fig. 4.5 Summer and winter isotherms as they might appear on a hypothetical continent. From *A Geography of Man*, 2d ed., by Preston E. James, p. 181. Copyright © 1959 by Ginn and Company. Redrawn by permission of John Wiley & Sons, Inc.

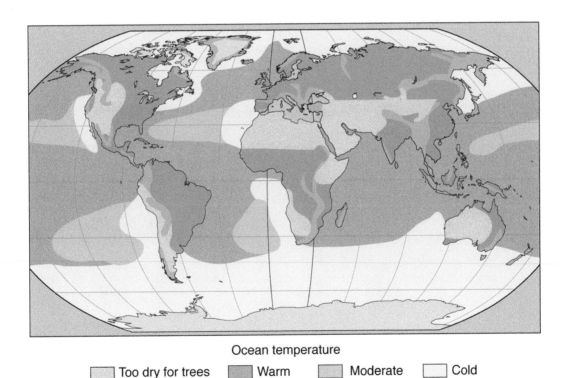

Ocean temperature

☐ Too dry for trees   ▨ Warm   ▨ Moderate   ☐ Cold

Fig. 4.6 Ocean temperatures determine, in part, the position of the dry zones on the continents. After Gerhard Schott; from *A Geography of Man*, 2d ed., by Preston E. James, p. 632. Copyright © 1959 by Ginn and Company. Redrawn by permission of John Wiley & Sons, Inc.

patterns. This is shown in Fig. 4.12, which compares locations in the Rocky Mountains. In the semiarid steppe climatic portion, the lowermost zone is a sagebrush basal plain; this is followed by a montane zone of Douglas-fir and spruce and fir. Above is the subalpine zone,

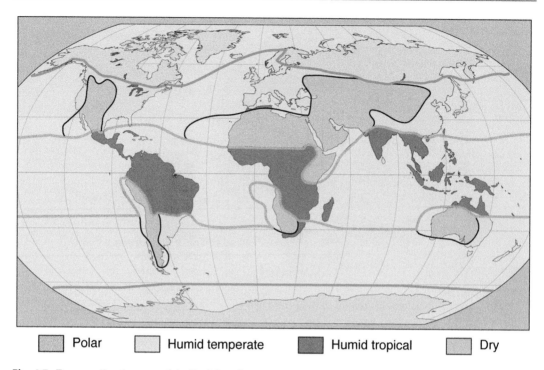

Fig. 4.7 Four ecoclimatic zones of the Earth based on temperature and moisture

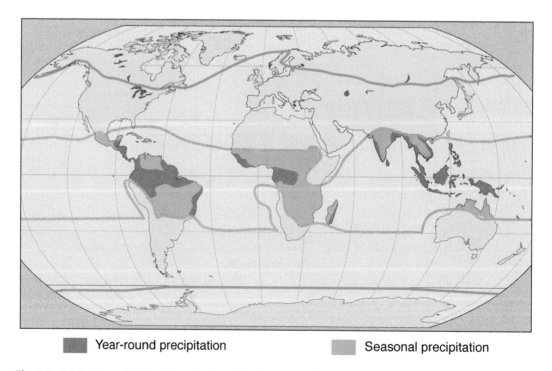

Fig. 4.8 Subdivisions of the humid tropical zone based on seasonal moisture distribution

Table 4.1 Broad plant formations and groups of climates

Formation	Köppen climate group
Tropical rain forest	A (Tropical rainy climates)
Tropical desert	B (Dry climates)
Temperate deciduous forest	C (Warm temperate climates)
Boreal forest	D (Snowy-forest climates)
Tundra	E (Polar climates)

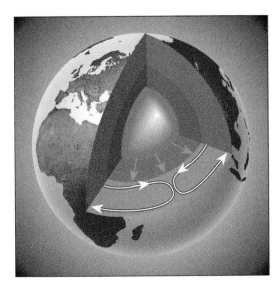

Fig. 4.9 The Earth's internal energy sources drive mantle convection and plate tectonics, causing mountain building. Base from Mountain High Maps. Copyright © by Digital Wisdom, Inc.

followed by alpine tundra, and then perennial ice and snow. This sequence of elevational zones repeats on mountain ranges throughout the lowland, semiarid climatic zone.

Orographically modified macroclimates, which exhibit elevational zonation, are referred to as **azonal** because they can occur in any ecoclimatic zone. The global distribution of the regions of the type is shown in Fig. 4.13.

4.2   Types of Ecoclimatic Zones, or Continental Ecoregions

The effects of latitude, continental position, and elevation combine to form the world's ecoclimatic zones, herein referred to as continental ecoregions. The geographic reasoning used in drawing boundaries of such regions involves several principles.

4.2.1   Principles of Ecological Climate Regionalization

1. *The boundaries of ecoclimate zones may coincide with certain climatic elements.* For example, the poleward limit of the **boreal** conifer zone corresponds roughly with the 10 °C (50 °F) isotherm for the warmest month. The 18 °C (64 °F) isotherm of the coolest month approximates the boundary of the poleward limit of plants that characterize the humid tropics, such as palms. *Generally speaking, however, a combination of factors rather than a single climatic factor determines such boundaries and this is more in line with the concept of ecological climate* (Bonan 2002).

2. *The climatic factors that delimit the main types of world vegetation must be treated as complex.* For example, to define "drought" temperature and precipitation must be treated together because drought is not just a matter of precipitation. A given amount of rainfall may produce a humid climate under temperate conditions, whereas the same amount of rainfall may produce arid conditions in hot zones. This leads to the situation that any given amount of precipitation may be interpreted differently across ecoregions: excessive precipitation in one area may be lacking in another.

3. *Specific major factors must be used to delimit the major ecoclimatic zones.* Examples would include the continuous high temperatures of the tropical lowlands, the deficiency of heat in

Fig. 4.10 Cloud
formations differentiating a
mountain boundary. View
west toward the Sangre de
Christo Mountains,
Colorado. Photograph by
Lev Ropes

Fig. 4.11 The elevation of
major mountain areas is
generally independent of
latitude

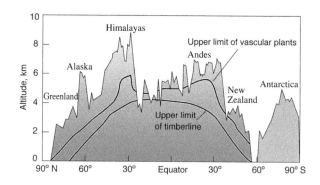

Fig. 4.12 Vertical
zonation in different
ecoclimatic zones along the
eastern slopes of the Rocky
Mountains in North
America. From
Schmithüsen (1976), p. 70

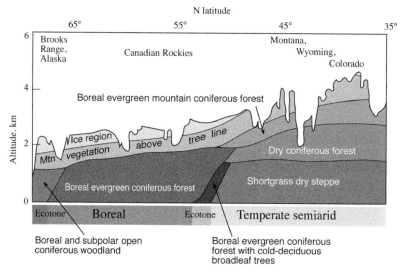

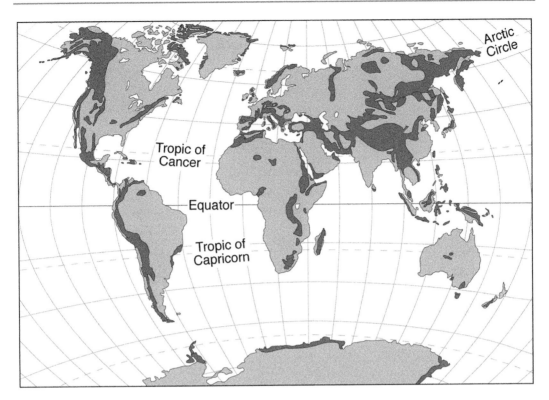

Fig. 4.13 The arrangement of orographically modified macroclimates, or highland climates, are distributed as a result of mountain building rather than latitude and therefore cut across other latitudinally based climates

the cold zone, and the lack of precipitation in the arid zone.

4. *Differences in climatic regime are used to define ecosystem boundaries* (Troll 1966). One can illustrate different climatic regimes by studying **climate diagrams**, or climographs (Walter et al. 1975). For example, tropical rainforest climates lack seasonal periodicity, whereas mid-latitude steppe climates have pronounced seasons. Climate diagrams show temperature and precipitation values but also the duration and intensity of relatively humid and relatively arid seasons, the severity of a cold winter, and the possibility of late or early frosts. With this information, it is possible to judge the climate from an ecological standpoint.

5. *Seasonality of precipitation affects the potential distribution of vegetation and must be considered in setting up ecoclimatic zones.* Within the arid zone, for example, deserts that receive only winter rain (Mojave Desert) can be distinguished from those that receive only summer rain (Chihuahuan Desert). Within the steppe zone, a semiarid steppe (short-grass prairie) climate that has a dry summer season and occasional drought can be distinguished from an arid semidesert (sagebrush shrubland) climate that has a very pronounced drought season plus a short humid season. The Great Plains grasslands are associated with the near absence of winter rains and the presence of spring rains.

6. *The delimitation of the ecological climates must incorporate the degree of dryness or of cold.* These two factors affect the distribution of vegetation. For example, the vegetation of the tropical savanna (wet–dry

tropics) is highly differentiated relative to variation in length of the dry season. A southern (coniferous) forest climate and northern (forest-tundra) climate can be distinguished within the Subarctic Division of the Polar Domain on the basis of temperature differences.

7. *The arrangement of ecoclimate zones depends largely on latitude and continental position with location-specific influences derived from elevation.* Mountain ranges that cut across latitudinally oriented ecoclimate zones support their own ecosystems. Elevation creates characteristic ecological zones that are variations of the lowland climate. Mountains show typical climatic characteristics depending on their location in the overall pattern of global ecoclimatic zones. The mountain ranges of Central America, for example, experience the same year-round, high-energy input and seasonal moisture regime consisting of relatively dry winters and rainy summers that are typical of their neighboring lowlands.

Beyond these principles, an additional point must be made. Many people think of climate as weather elements (e.g., mean annual temperature or precipitation) that go into a database or constitute a data layer in a **geographic information system** (GIS). Ecological climate is emphatically more than that. It is an integrated concept and consideration should be given to how climate affects all other elements of the ecosystem, both at a site and across the landscape.

The most frequently used climate classification that follows these principles was developed and proposed by Wladimir Köppen in his seminal book on the subject (Köppen 1931). Building on ideas of an earlier bioclimatologist, de Candolle (1874), Köppen carefully observed the climatic conditions required for the growth of various groups of plants, and he related variations in vegetation to temperature and precipitation. Köppen's system is based on the concept that native vegetation is the best expression of climate. Perhaps the main shortcoming of this classification lies in the fact that

the boundaries of certain climate types do not correspond with the observed boundaries of natural landscapes. This led Trewartha (1968) to slightly modify the Köppen scheme by establishing more realistic criteria to distinguish climate types.

I developed an approach to mapping ecoregions for the United States (Bailey 1976, revised 1994), North America (Bailey 1998), and the world's continents (Bailey 1989) based on these principles. The boundaries of these ecoregions follow the subdivision of the Earth into climatic zones established by Köppen as modified by Trewartha (1968, Table 4.2),[2] although some modifications have been made to maximize correspondence of the regions with the vegetation (plant formation classes).

4.2.2 The Köppen–Trewartha Classification of Climates

The Köppen–Trewartha classification identified six main groups of climate, and all but one—the dry group—are thermally defined. They are as follows:

Based on temperature criteria

A. Tropical: Frost limit in continental locations; in marine areas 18 °C for the coolest month

C. Subtropical: 8 months 10 °C of above

D. Temperate: 4 months 10 °C or above

E. Boreal: 1 (warmest) month 10 °C or above

F. Polar: All months below 10 °C

Based on precipitation criteria

B. Dry: Outer limits, where potential evaporation equals precipitation

These groups are subdivided into 15 types based on seasonality of precipitation or on degree of dryness or cold. They range from the icecaps

---

[2] Other methods for mapping zones at the global scale are those of Thornthwaite (1931, 1933), Holdridge (1947), and Walter and Box (1976). All methods appear to work better in some areas than in others and to have gained their own adherents. The Köppen system was chosen as the basis for ecoregion delineation because it has become the international standard for geographical purposes.

Table 4.2 Regional climates[a]

Köppen group and types	Ecoregion equivalents
A Tropical and humid climates	**Humid tropical domain** (400)
Tropical wet (Ar)	Rainforest division (420)
Tropical wet-dry (Aw)	Savanna division (410)
B Dry climates	**Dry domain** (300)
Tropical/subtropical semiarid (BSh)	Tropical/subtropical steppe division (310)
Tropical/subtropical arid (BWh)	Tropical/subtropical desert division (320)
Temperate semiarid (BSk)	Temperate steppe division (330)
Temperate arid (BWk)	Temperate desert division (340)
C Subtropical climates	**Humid temperate domain** (200)
Subtropical dry summer (Cs)	Mediterranean division (260)
Humid subtropical (Cf)	Subtropical division (230)
	Prairie division (250)[b]
D Temperate climates	
Temperate oceanic (Do)	Marine division (240)
Temperate continental, warm summer (Dca)	Hot continental division (220)
	Prairie division (250)[b]
Temperate continental, cool summer (Dcb)	Warm continental division (210)
	Prairie division (250)[b]
E Boreal climates	**Polar domain** (100)
Subarctic (E)	Subarctic division (130)
F Polar climates	Tundra division (120)
Tundra (Ft)	
Ice Cap (Fi)	
Definitions and Boundaries of the Köppen–Trewartha System	
Ar	All months above 18 °C and no dry season
Aw	Same as Ar, but with 2 months dry in winter
BSh	Potential evaporation exceeds precipitation, and all months above 0 °C
BWh	One-half the precipitation of BSh, and all months above 0 °C
BSk	Same as BSh, but with at least 1 month below 0 °C
BWk	Same as BWh, but with at least 1 month below 0 °C
Cs	8 months 10 °C, coldest month below 18 °C, and summer dry
Cf	Same as Cs, but no dry season
Do	4–7 months above 10 °C, coldest month above 0 °C
Dca	4–7 months above 10 °C, coldest month below 0 °C, and warmest month above 22 °C
Dcb	Same as Dca, but warmest month below 22 °C
E	Up to 3 months above 10 °C
Ft	All months below 10 °C
Fi	All months below 0 °C
A/C boundary = Equatorial limits of frost; in marine locations, the isotherm of 18 °C for coolest month	
C/D boundary = 8 months 10 °C	
D/E boundary = 4 months 10 °C	
E/F boundary = 10 °C for warmest month	
B/A, B/C, B/D, B/E boundary = Potential evaporation equals precipitation	

[a]Based on the Köppen system of classification (Köppen 1931), as modified by G.T. Trewartha (1968) and Trewartha et al. (1967)

[b]Köppen did not recognize the Prairie as a distinct climatic type. The ecoregion classification system represents it at the arid sides of the Cf, Dca, and Dcb types

at high latitudes, to the tropical wet climate at low latitudes.

Trewartha (1968) describes them briefly. The low latitudes contain a winterless, frostless belt with adequate rainfall. This is the tropical humid climate, or the *A* group. It is subdivided into two types, tropical wet (*Ar*) and tropical wet-and-dry (*Aw*). On the low-latitude margins of the middle latitudes, where winters are mild and killing frosts only occasional, is the subtropical belt, or *C* group. Here two subdivisions are recognized: subtropical dry summer (*Cs*) and subtropical humid (*Cf*). Poleward from the subtropics is the temperate belt, or *D* group. It contains two types, temperate continental (*Dc*) and temperate oceanic (*Do*). Two subtypes of temperate continental are recognized: a more moderate one with hot summers and cold winters (*Dca*), and a more severe subtype (*Dcb*), located poleward, which has warm summers and rigorous winters. Still farther poleward is the boreal of subarctic belt, the *E* group. It has not been subdivided. In the very high latitudes are the summerless, polar climates (*F* group), subdivided into the tundra climate (*Ft*) and ice-cap climate (*Fi*). The dry climates, group *B*, are subdivided into semiarid or steppe type (*BS*) and an arid or desert type (*BW*). A further subdivision separates the hot tropical–subtropical deserts and steppes (*BWh*, *BSh*) from the cold temperate–boreal deserts and steppes (*BWk*, *BSk*) of middle latitudes.

Highland climates, which are low-temperature variants of climates at low elevations in similar latitudes, are designated by the letter *H*.

4.3   Distribution of the Continental Regions

Although we can define ecoregions climatically, they are most effectively treated by combining and rearranging the 15 climatic types to maximize correspondence with major plant formations. Through this process I have mapped

the Earth into zones called *ecoregion provinces*, each of which has characteristic ecosystems. This map is based on a world map of natural climate–vegetation landscape types in the *Fiziko-geograficheskii atlas mira* (Gerasimov 1964). Eight-six such subdivisions are recognized (Bailey 1989), but for simplification I have grouped them into 15 *divisions*. They range from the tundra at high latitudes to the rainforest at low latitudes. Mountains exhibiting **elevational zonation** and having the climatic regime of the adjacent lowlands are distinguished according to the character of the zonation.

We can further simplify this classification of ecosystems by grouping the divisions into four large regions called *domains*. Four such subdivision are recognized—three are humid zones, thermally differentiated: *polar*, with no warm season; *humid temperate*, rainy with mild to severe winters; *humid tropical*, rainy with no winters. The fourth, *dry*, is defined on the basis of moisture alone, and transects the otherwise humid domains.

These domains can be described briefly. In the very high latitudes lie the polar domain, differentiated on the basis of ice formation and plant development into icecap, tundra, and subarctic tayga divisions. In the mid latitudes is the humid temperate domain of mid-latitude forests, differentiated according to the importance of winter frost into warm continental, hot continental, subtropical, marine, prairie, and mediterranean divisions. In low and middle latitudes is the dry domain, differentiated on the basis of rainfall (steppe versus desert), and winter temperature (cold versus warm), into tropical/subtropical steppe, tropical/subtropical desert, temperate steppe, and temperate desert. In the low latitudes lie the humid tropical domain, differentiated on the basis of rainfall seasonality into savanna and rainforest divisions.

We have identified two principles in the global pattern of ecosystem regions. One is the principle of regularity which relates to those features of the Earth's surface that are associated with climate; the other is the principle of

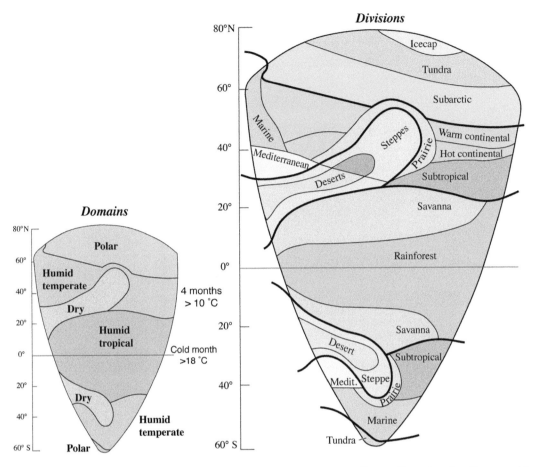

Fig. 4.14   Patterns of ecoregions that might occur on a hypothetical continent of low, uniform elevation. Compare with map of continents, Plate 2

irregularity which relates to all those features associated with land surface forms. The principle of regularity allows us to forecast the kinds of associated features that can be expected at any given latitude and longitudinal position. However, irregularities distort these regular patterns on each specific continent.

We can plot the regular pattern of these regions over the Earth's continents—a pattern that is generalized in Fig. 4.14. Because the arrangement of these regions is regular, it is possible to predict the kind of region that will be found in any particular part of the Earth's land areas. Mountains provide the distortions observed on the actual map of the ecoregions in Plate 2 (Maps, p. 165). Mountains exhibiting elevational zonation and having the climatic

regime of the adjacent lowlands do not appear on our hypothetical continent because they are unique for each land mass.

Each group includes regions in different parts of the world that are broadly similar with respect to surface features and the cover of vegetation. For example, the tropical/subtropical steppe division (310) of the dry domain is found on all continents. These steppes typically are grasslands of short grasses and other herbs with local shrub and woodland. Pinyon-juniper woodland, for example, grows on the Colorado Plateau of the United States (Fig. 4.15). These areas may support limited grazing, but are not moist enough for cultivated crops without irrigation.

Table 4.3 lists climate, vegetation, and soil types associated with each zone (division). In

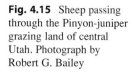
Fig. 4.15 Sheep passing through the Pinyon-juniper grazing land of central Utah. Photograph by Robert G. Bailey

Chaps. 5–8, we describe the characteristics of each of the ecoregion domains and divisions of the continents. These characteristics result from the interplay of surface features, climate, vegetation, soil, water, and fauna. The proportion of the Earth's land area included in each group is shown in Table 4.4.

The descriptions are based on information compiled and summarized from many sources, the most important of which are James (1959), Hidore (1974), Strahler and Strahler (1989), Walter (1984), and Schultz (1995). Climate descriptions are based largely on the Köppen–Trewartha classification, but we also explain them in terms of air masses and frontal zones which are shown in Appendix A. In presenting the climate, we make use of climate diagrams of representative stations adapted from the well-known system of Heinrich Walter (Walter and Lieth 1960–1967; Walter et al. 1975). Soil information is founded on the soil classification established by the USDA Soil Conservation Service (U.S. Department of Agriculture 1938; U.S. Department of Agriculture and Soil Survey Staff 1975), now the Natural Resources Conservation Service.

4.4   Relationship to Other Ecoregional Zoning Systems

Several different approaches for defining ecoregions have grown out of the system presented in this book (for details, see Appendix D). One example would be The Nature Conservancy (TNC), which has shifted its focus from conservation of single species and small sites to conservation planning on an ecoregional basis (The Nature Conservancy 1997). On their map, this text appears:

> This map was developed as a coordinated effort by TNC US ecoregional planning teams. Based on Bailey 1994, the ecoregions have been modified for both biological and administrative purposes.

They included other criteria, such as the distribution of species as well as the location of conservation units within ecoregions to be administered or managed. According to Denny Grossman of The Nature Conservancy (personal communication), land use may have been included in the mix of factors they considered.

Hargrove and Luxmore (1998) created ecoregions of the United States by applying a

Table 4.3 General environmental conditions for ecoregion divisions

Name of division	Equivalent Köppen–Trewartha climates	Zonal vegetation	Zonal soil type[a]
110 Icecap			
120 Tundra	Ft	Ice and stony deserts: tundras	Tundra humus soils with soilifluction (Entisols, Inceptisols, and associated Histosols)
130 Subarctic	E	Forest-tundras and open woodlands; tayga	Podzolic (Spodosols and Histosols)
210 Warm continental	Dcb	Mixed deciduous–coniferous forests	Gray-brown podzolic (Alfisols)
220 Hot continental	Dca	Broadleaved forests	Gray-brown podzolic (Alfisols)
230 Subtropical	Cf	Broadleaved–coniferous evergreen forests; coniferous–broadleaved semi-evergreen forests	Red-yellow podzolic (Ultisols)
240 Marine	Do	Mixed forests	Brown forest and gray-brown podzolic (Alfisols)
250 Prairie	Cf, Dca, Dcb	Forest-steppes and prairies; savannas	Prairie soils, Chernozems (Molisols)
260 Mediterranean	Cs	Dry steppe; hardleaved evergreen forests, open woodlands and shrub	Soils typical of semiarid climates associated with grasslands
310 Tropical/subtropical steppe	BSh	Open woodland and semideserts; steppes	Chestnut, brown soils, and sierozems (Mollisols, Aridisols)
320 Tropical/subtropical desert	BWh	Semideserts; deserts	Desert (Aridisols)
330 Temperate steppe	BSk	Steppes; dry steppes	(same as BSh)
340 Temperate desert	BWk	Semideserts and deserts	(same as BWh)
410 Savanna	Aw, Am	Open woodlands, shrubs and savannas; semievergreen forest	Latisols (Oxisols)
420 Rainforest	Ar	Evergreen tropical rain forest (selva)	Latisols (Oxisols)

[a]Great soil group. Names in parenthesis are Soil Taxonomy soil orders (U.S. Department of Agriculture, Soil Survey Staff 1975). Described in the Glossary, p. 153

clustering algorithm to factors derived from climate and soils data. Although the variables were chosen because of their hypothesized relationship to vegetation patterns, this relationship was not formally defined.

A similar approach to ecoregionalization is to overlay thematic maps either manually (Omernik 1987; Gallant et al. 1995; Harding et al. 1997; Olson et al. 2001) or within a GIS (Host et al. 1996; Bernert et al. 1997; Zhou et al. 2003). Typically, ecoregion boundaries are placed where boundaries of several input layers are located in close proximity to one another. Boundaries defined in this manner represent areas of spatially abrupt changes in ecological characteristics. The overlay method has been referred to as the weight of evidence approach (McMahon et al. 2001) and the gestalt method (Bailey 2009; Jepson and Whittaker 2002) because homogeneous-appearing regions are recognized and boundaries are drawn through a process of intuitive and holistic reasoning based largely on visual appearance. Ambiguity arises from the fact that boundaries of input layers

Table 4.4 Approximate area and proportionate extent of ecoregions

	km^2	Percent
100 Polar domain	38,038,000	26.00
110 Icecap division	12,823,000	8.77
M110 Icecap regime mts	1,346,000	0.92
120 Tundra division	4,123,000	2.82
M120 Tundra regime mts	1,675,000	1.14
130 Subarctic division	12,259,000	8.38
M130 Subarctic regime mts	5,812,000	3.97
200 Humid temperate domain	22,455,000	15.35
210 Warm continental division	2,187,000	1.49
M210 Warm continental regime mts	1,135,000	0.78
220 Hot continental division	1,670,000	1.14
M220 Hot continental regime mts	485,000	0.33
230 Subtropical division	3,568,000	2.44
M230 Subtropical regime mts	1,543,000	1.05
240 Marine division	1,347,000	0.92
M240 Marine regime mts	2,194,000	1.50
250 Prairie division	4,419,000	3.02
M250 Prairie regime mts	1,256,000	0.88
260 Mediterranean division	1,090,000	0.75
M260 Mediterranean regime mts	1,561,000	1.07
300 Dry domain	46,806,000	32.00
310 Trop/sub steppe division	9,838,000	6.73
M310 Trop/sub steppe regime mts	4,555,000	3.11
320 Trop/sub desert division	17,267,000	11.80
M320 Trop/sub desert regime mts	3,199,000	2.19
330 Temperate steppe division	1,790,000	1.22
M330 Temperate steppe regime mts	1,066,000	0.73
340 Temperate desert division	5,488,000	3.75
M349 Temperate desert regime mts	613,000	0.42
400 Humid tropical domain	38,973,000	26.64
410 Savanna division	20,641,000	14.11
M410 Savanna regime mts	4,488,000	3.07
420 Rainforest division	10,403,000	7.11
M420 Rainforest regime mts	3,440,000	2.35

Source: World Conservation Monitoring Centre

rarely conform to one another (the "GIS trap"; see Bailey 1988). The general procedure is consistent in concept with the empirical approach because it emphasizes pattern over process (Omernik 2004).

The empirical approach seeks to discern patterns in the data. The resultant maps frequently show highly fragmented map units, with small, noncontiguous units of the same region distributed over wide areas. In the approach presented in this book, the boundaries of terrestrial ecoregions are determined to a considerable extent by the changing nature of the climate over large areas. This approach takes into account compensating factors (see Chap. 11) that override the climatic effect. For example, in the High Plains of the semiarid southwestern United States, forests extend along streams because of the extra water supply. Ponderosa pine and shrub islands within the grasslands of these regions indicate rocky soil conditions, forming reservoirs of water for taproots. They occur there because of the ground water condition; not because of the climate. In these cases, a map of climate-based ecoregions ignores such areas and relegates **edaphically** controlled ecosystems to a lower level of classification and more detailed maps. The emphasis in this kind of mapping is on understanding pattern in term of process.

Regarding land use, the ecoregion delineations described in this book are based solely on biophysical factors (i.e., climate, landform, and vegetation). This is an important difference from the systems described above in which land use is one of the delineators. One of the major advantages of the biophysical approach (as opposed to directly mapping land cover) is its ability to predict the potential character of sites where natural ecosystems have been profoundly modified (e.g., by land clearance or fire) or replaced by exotic species regarded as pests and weeds. Because land use has not been considered in drawing up regional boundaries, exact correspondence may not occur between these biophysical-based ecoregions and existing vegetation patterns.

The system in this book also stands in contrast to other systems that use the distribution limits of species and races of plants and animals as criteria for determining the boundaries of ecoregions. Sometimes, the distribution limits for several species might coincide with an ecoregion boundary if that boundary follows some barrier that prevents range expansion, such as where plains meet mountains. Often, however, the range of a species does not stop abruptly at the border of an ecoregion but continues for a distance into the adjacent ecoregion. The reason for this seems to be that some isolated areas of suitable habitat usually occur in the adjacent region. Furthermore, because, at small map scales, **physiognomy** (life-form) is the best expression of ecological conditions (Küchler 1973; Gosz and Sharpe 1989), floristic and faunistic differences are best left to maps with other purposes.

Because physiognomy is basic and applicable without exception anywhere on the Earth, it was selected to serve as the source for the criteria necessary to establish the basis for regional differences. These criteria permit a uniform approach throughout the world and put the various parts of the world on a comparable basis.

In this book, no attempt is made to develop separate systems to define freshwater and terrestrial ecosystem units. In some other systems, there is a false dichotomy being drawn between aquatic and terrestrial systems. Ecosystems are not aquatic or terrestrial; they are geographic system that include both aquatic and terrestrial components that are integrated, or combined, in a particular geographic area. The streams and rivers that drain an ecological unit have common biological and physical characteristics that are controlled by the nature of the ecological unit they are embedded in (Omernik and Bailey 1997; Bailey 2009).

Finally, the boundaries of ecoregions on some maps are very irregular; thus implying accuracy and precision. In contrast, the view of ecoregions promulgated in this book looks on the boundaries between ecoregions as representing climatic gradients that reflect gradual change and tend to be indistinct. Such boundaries cannot be located precisely. The complex boundary lines on ecoregion maps give a false sense of accuracy and precision. More details are presented in Chap. 11 and Bailey (2004).

References

Bailey RG (1976) Map: ecoregions of the United States. USDA Forest Service, Intermountain Region, Ogden, UT, Scale 1:7,500,000

Bailey RG (1988) Problems with using overlay mapping for planning and their implications for geographic information systems. Environ Manage 12:11–17

Bailey RG (1989) Explanatory supplement to ecoregions map of the continents. Environ Conserv 16:307–309, with separate map at 1:30,000,000

Bailey RG (1994) Map: ecoregions of the United States (rev.). USDA Forest Service, Washington, DC, Scale 1:7,500,000

Bailey RG (1998) Ecoregions map of North America: explanatory note. Miscellaneous Publication 1548. USDA Forest Service, Washington, DC, 10 pp, with separate map at 1:15,000,000

Bailey RG (2004) Identifying ecoregion boundaries. Environ Manage 34(Suppl 1):S14–S26

Bailey RG (2009) Ecosystem geography: from ecoregions to sites, 2nd edn. Springer, New York, NY, 251 pp

Bernert JA, Eilers JM, Sullivan TJ, Freemark KE, Ribic C (1997) A quantitative method for delineating regions: an example from the Western Corn Belts Ecoregions of the USA. Environ Manage 21:405–420

Bonan GB (2002) Ecological climatology: concepts and applications. Cambridge University Press, New York, NY, 678 pp

de Candolle APA (1874) Constitution Dans le Regne Vegetal de Groupes Physiologiques Applicables a la Geographie Ancienne et Moderne. Archives des Science Physiques et Naturelles, Geneva

Gallant AL, Binnian EF, Omernik JM, Shasby MB (1995) Ecoregions of Alaska. Professional Paper 1567. US Geological Survey, Washington, DC, with separate map at 1:5,000,000

Gerasimov IP (ed) (1964) Types of natural landscapes of the earth's land areas. Plate 75. In: Fiziko-geograficheskii atlas mira (physico-geographic atlas of the world). USSR Academy of Science and Main Administration of Geodesy and Cartography, Moscow. Scale 1:80,000,000

Gosz JR, Sharpe PJH (1989) Broad-scale concepts for interactions of climate, topography, and biota at biome transitions. Landsc Ecol 3:229–243

Harding JS, Winterbourn MJ, McDiffett WF (1997) Stream fauna and ecoregions in South Island, New Zealand: do they correspond? Arch Hydrobiol 140 (3):289–307

Hargrove WW, Luxmore RJ (1998) A clustering technique for the generation of customizable ecoregions. In: Proceedings ESRI Arc/INFO users conference. http://proceedings.esri.com/library/userconf/proc97/proc97/TO250/pap226/p226.htm

Hidore JJ (1974) Physical geography: earth systems. Scott, Foresman and Co., Glenview, IL, 418 pp

Holdridge LR (1947) Determination of world plant formations from simple climatic data. Science 105:367–368

Host GE, Polzer PL, Mladenoff DJ, White MA, Crow TR (1996) A quantitative approach to developing regional ecosystem classifications. Ecol Appl 6:608–818

James PE (1959) A geography of man, 2nd edn. Ginn, Boston, MA, 656 pp

Jepson P, Whittaker RJ (2002) Ecoregions in context: a critique with special reference to Indonesia. Conserv Biol 16:42–57

Köppen W (1931) Grundriss der Klimakunde. Walter de Gruyter, Berlin, 388 pp

Küchler AW (1964) Potential natural vegetation of the conterminous United States. Special Publication 36.

American Geographical Society, New York, NY, 116 pp, with separate map at 1:3,168,000

Küchler AW (1973) Problems in classifying and mapping vegetation for ecological regionalization. Ecology 54:512–523

McMahon G, Gregonis SM, Waltman SW, Omernik JM et al (2001) Developing a spatial framework of common ecological regions for the conterminous United States. Environ Manage 28:293–316

The Nature Conservancy (1997) Designing a geography of hope: guidelines for ecoregion-based conservation in The Nature Conservancy. The Nature Conservancy, Arlington, VA, 84 pp

Olson DM, Dinerstein E, Wikramanayake ED, Burgess ND et al (2001) Terrestrial ecoregions of the world: a new map of life on earth. Bioscience 51(11):933–938

Omernik JM (1987) Ecoregions of the conterminous United States (map supplement). Ann Assoc Am Geogr 77:118–125

Omernik JM (2004) Perspectives on the nature and definition of ecological regions. Environ Manage 34(S1): S27–S38

Omernik JM, Bailey RG (1997) Distinguishing between watersheds and ecoregions. J Am Water Resour Assoc 33:1–15

Schmithüsen J (1976) Atlas zur biogeographie. Bibliographisches Institut, Mannheim-Wien-Zurich, 88 pp

Schultz J (1995) The ecozones of the world: the ecological divisions of the geosphere (trans: from German by I. and D Jordan). Springer, Heidelberg, 449 pp

Strahler AN, Strahler AH (1989) Elements of physical geography, 4th edn. Wiley, New York, NY, 562 pp

Thornthwaite CW (1931) The climates of North America according to a new classification. Geogr Rev 21:633–655, with separate map at 1:20,000,000

Thornthwaite CW (1933) The climates of the earth. Geogr Rev 23:433–440, with separate map at 1:77,000,000

Trewartha GT (1968) An introduction to climate, 4th edn. McGraw-Hill, New York, NY, 408 pp

Trewartha GT, Robinson AH, Hammond EH (1967) Physical elements of geography, 5th edn. McGraw-Hill, New York, NY, 527 pp

Troll C (1966) Seasonal climates of the earth. The seasonal course of natural phenomena in the different climatic zones of the earth. In: Rodenwaldt E, Jusatz HJ (eds) World maps of climatology, 3rd edn. Springer, Berlin, pp 19–28, with separate map at 1:45,000,000 by C. Troll and K.H. Paffen

U.S. Department of Agriculture, (1938) Soils and men, 1938 yearbook of agriculture. U.S. Government Printing Office, Washington, DC, 1232 pp

U.S. Department of Agriculture, Soil Survey Staff (1975) Soil taxonomy: a basic system for making and interpreting soil surveys. Agriculture handbook 436. U.S. Department of Agriculture, Washington, DC, 754 pp

Walter H (1984) Vegetation of the earth and ecological systems of the geo-biosphere [trans: from German by

Muise O], 3rd rev enlarged edn. Springer, Berlin, 318 pp

Walter H, Box E (1976) Global classification of natural terrestrial ecosystems. Vegetatio 32:75–81

Walter H, Harnickell E, Mueller-Dombois D (1975) Climate-diagram maps of the individual continents and the ecological climatic regions of the earth. Springer, Berlin, 36 pp, with 9 maps

Walter H, Lieth H (1960–1967) Klimadiagramm weltatlas. Jena: G. Fischer. Maps, diagrams, profiles. Irregular pagination.

Zhou Y, Narumalani S, Waltman WJ, Waltman SW, Palecki MA (2003) A GIS-based spatial pattern analysis model for eco-region mapping and characterization. Int J Geogr Inf Sci 17:445–462

Ecoregions of the Continents: The Polar Ecoregions

5

5.1   100 Polar Domain

Polar and Arctic air masses chiefly control climates of the polar domain, located at high latitudes. With the exception of the ice cap climates, they lie entirely in the Northern Hemisphere. In general, climates in the polar domain have low temperatures, severe winters, and small amounts of precipitation, most of which falls in summer. Polar systems are dominated by a periodic fluctuation of solar energy and temperature, in which the annual range is far greater than the diurnal range (see Fig. 4.2, p. 29). This contrasts with the tropics where the major periodic fluctuation is the diurnal one, and the mid-latitude systems, as we will see, are subject to fluctuations in both annual and diurnal energy patterns.

The intensity of the solar radiation is never very high compared to ecosystems of the middle latitudes and tropics. On the poleward margins, although summer insolation persists for many hours, temperatures do not get very high because the intensity is low and because much of the energy goes to evaporate water and melt snow or ice. More energy is given off by terrestrial radiation than is received from solar radiation. In order for this situation to persist, a supplementary heat source must exist to provide the difference. This supplementary energy source is heat carried poleward by wind and water currents. It maintains the Arctic temperatures at a level much higher than would solar radiation alone.

The high-latitude climates have low annual total evaporation, always less than 50 cm, reflecting the prevailing low air and soil temperatures. The frozen condition of the soil in several consecutive winter months causes plant growth to cease. Snow that falls in this period is retained in surface storage until spring thaw releases it for infiltration and runoff. The growing season for crops is short in the subarctic zone, but low air and soil temperatures are partly compensated for by great increase in day length.

In areas where summers are short and temperatures are generally low throughout the year, thermal efficiency, rather than effectiveness of precipitation, is the critical factor in plant distribution and soil development. Three major regional divisions have been recognized and delimited in terms of thermal efficiency—the *icecap*, *tundra*, and the *subarctic* (tayga). The world map of the polar ecoregions of the continents (Fig. 5.1) shows the locations of these ecoregions. **Climate diagrams** in Fig. 5.2 provide general information on the character of the climate in two of these divisions.

These climate diagrams express differences in the climatic regime of the divisions within the domains by comparing annual temperature and moisture cycles. A relatively dry season is depicted by the precipitation curve falling below the temperature curve. The location of the weather station and its altitude, as well as the average annual temperature and precipitation, are shown above the graphs.

R.G. Bailey, *Ecoregions*, DOI 10.1007/978-1-4939-0524-9_5, © Springer Science+Media, LLC 2014

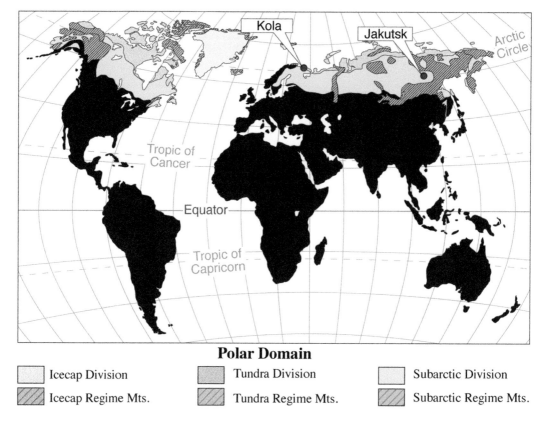

Polar Domain

Icecap Division	Tundra Division	Subarctic Division
Icecap Regime Mts.	Tundra Regime Mts.	Subarctic Regime Mts.

Fig. 5.1 Divisions of the continental polar domain

5.2   110 Icecap Division

These are the ice sheets of Greenland, Ellesmere Island, Baffin Island, and Antarctica. Mean annual temperature is much lower than that of any other climate, with no month above freezing, defining this climate as *Fi*. Precipitation, almost all occurring as snow, is small, but accumulates because of the continuous cold. Driving blizzard winds are frequent. These regions are subjected to long periods of darkness and light.

Because of low monthly mean temperatures throughout the year over the ice sheets, this environment is almost devoid of vegetation and soils (Fig. 5.3). Only a few plant species can survive the Antarctic climate. The dominant plants are algae, lichens, and mosses—plants adapted to limited water supply and scant nutrients and soil. The few species of animals found on the ice margins are associated with a marine habitat.

In the Antarctic regions, because of the absence of predatory animals, the flightless penguins find conditions on the coasts and on the isolated islands of the southern oceans ideal as breeding places.

Terrestrial invertebrates include a few species such as nematodes and springtails. Some starve and dehydrate themselves during cold weather because any water left in their bodies would encourage deadly ice crystals to form. Others produce antifreeze chemicals.

5.3   120 Tundra Division

The northern continental fringes of North America, Iceland and Spitsbergen, coastal Greenland, the Arctic coast of Eurasia from the Arctic Circle northward to about the 75th parallel, lie within the outer zone of control of Arctic air masses. This produces the tundra climate that Trewartha

Fig. 5.2 Climate diagrams from the tundra of northern Russia and from the extremely cold, continental boreal (tayga) regions of eastern Siberia. Redrawn from Walter et al. (1975)

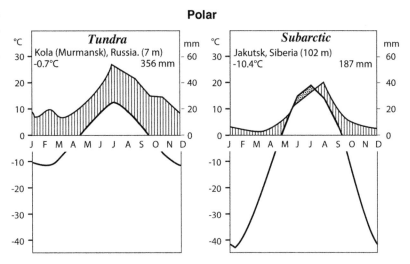

Polar

Fig. 5.3 A glacier and mountains in Antarctica. Taken during the Byrd Expedition to South Pole, 1946–47. Image # 123435, American Museum of Natural History, Library

(1968) designated by symbol *Ft*. Average temperature of the warmest month lies between 10 and 0 °C. The tundra regions occupy some 5 % of the land surface of the Earth.

The tundra climate has very short, cool summers and long, severe winters (see Fig. 5.2, climate diagram for Kola [near Murmansk], Russia). No more than 188 days per year, and sometimes as few as 55, have a mean temperature higher than 0 °C. Annual precipitation is light, often less than 200 mm, but because potential evaporation is also very low, the climate is humid. However, because precipitation is usually in the form of snow and ice, animals and plants cannot use it.

Tundra is the characteristic vegetation of the Polar Regions. Three chief types are recognized: these are *grass tundra*, *brush tundra*, and the *desert tundra*. Vegetation in the central part is grass tundra, a treeless plain of low-growing plants adapted to the climate's low temperatures, short growing season, and low precipitation. It consists of grasses, sedges, and lichens, with willow shrubs (Fig. 5.4). As in the Antarctic, mosses and lichens flourish because they can tolerate the freezing temperatures. Farther south, the vegetation changes into brush tundra, or birch–lichen woodland, then into a needleleaf forest. In some places, a distinct tree line separates the forest from tundra. Köppen (1931) used this line, which coincides approximately with the 10 °C isotherm of the warmest month, as a boundary between subarctic and tundra climates. Farther poleward the tundra breaks up into detached "oases" in the sheltered hollows, separated by expanses of bare rock or **regolith**. This is the desert tundra. The arctic flora is poor in plant species; only a few hundred species grow in the entire Arctic, compared to over 100,000 in the tropics.

In contrast to surfaces exposed to tropical rainy climate, soil particles of tundra derive almost entirely from mechanical breakup of the parent rock, by continual freezing and thawing, with little or no chemical alteration. **Tundra soils** (Entisols, Inceptisols, and associated

Fig. 5.4 Grass tundra in Alaska. Photograph by USDA Forest Service

Histosols), with weakly differentiated horizons, dominate. As in the northern continental interior, the tundra has a permanently frozen sublayer of soil known as **permafrost** (Fig. 5.5).[1] The permafrost layer is more than 300 m thick throughout the region; seasonal thaw reaches only 10–60 cm below the surface.

Probably half of the tundra surface is covered by water in the summer and ice in the winter. Poor drainage is a dominant characteristic. Permafrost prevents the percolation of meltwater into the regolith. The continental ice sheets that repeatedly scoured these areas left a rolling topography relatively free of weathered material. The depressions are filled with lakes and swamps, and bogs are everywhere. The streams that exist meander extensively on the surface, from depression to depression, or swamp to swamp. The large rivers that cross the tundra are **exotic rivers** and subject to spring flooding. Since these streams have their headwaters equatorward of the mouth, the spring thaw takes place first upstream. The meltwater starts downstream, only to encounter a channel blocked by a still-frozen stream. The only outlet for the meltwater is out of the channel over the ice and surrounding

land. In this fashion, immense areas of the Arctic tundra are flooded in the spring.

One aspect of the hydrologic cycle in the Arctic warrants particular notice. Whereas in most parts of the world the change of water to water vapor is a primary means of heating the atmosphere, in the Arctic this process is greatly reduced. Outside the Arctic, when condensation occurs and the water vapor is returned to a liquid or solid state, the energy utilized in evaporating water is again released into the atmosphere and heats it.

Geomorphic processes are distinctive in the tundra, resulting in a variety of curious landforms. Under a protective layer of sod, water in the soil melts in summer to produce a thick mud that sometimes flows downslope to create bulges, terraces, and lobes on hillsides. The freeze and thaw of water in the soil also sorts the coarse particles from the fine particles, giving rise to such patterns in the ground as rings, polygons, and stripes made of stone. The coastal plains have numerous lakes of **thermokarst** origin, formed by melting groundwater.

The richness and variety of the fauna of the Arctic regions are remarkable. Many of the land animals of the tayga migrate northward onto the tundra during the summer months. The reindeer of Eurasia and the similar, but smaller, caribou of North America are the most important of these. Another large herbivore, the musk oxen, occurs in isolated herds, grazing on the more luxuriant patches of desert tundra. Two small herbivora— the Arctic hare and the lemming—are also widespread.

Following the herds are a number of predatory animals, such as wolf and fox. The drift ice of the polar sea or the immediate coastal margins is the habitat of the polar bear, preying chiefly on marine life, such as the seal and the walrus.

Birds and insects are particularly numerous. Mosquitoes are probably present in greater numbers in summer over the tundra than anywhere else on the Earth. Flies are abundant, especially near human settlements. Attracted in part by this abundance of insect life, many migrating birds make the tundra their goal in summer. The boggy

---

[1] This map is highly generalized. For a more detailed presentation, see Brown et al. (1997).

Fig. 5.5 Distribution of permafrost in the Northern Hemisphere. From *Glacial and Quaternary Geology* by R.F. Flint, p. 270. Copyright © 1971 by John Wiley & Sons, Inc. Reprinted by John Wiley & Sons, Inc.

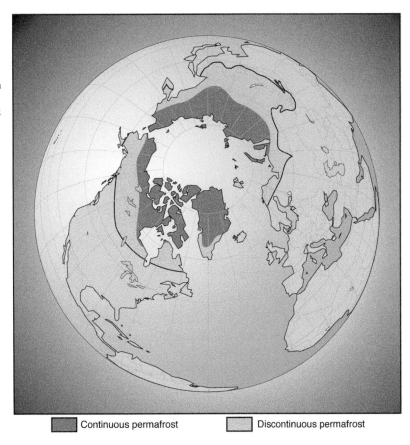

Continuous permafrost   Discontinuous permafrost

tundra offers an ideal summer environment for waterfowl, sandpipers, and plovers.

Sea life is rich. Although reptiles, amphibians, and freshwater fish are absent, the inhabitants of the polar oceans are varied and numerous.

Direct human impact in the tundra regions has been minimal until recent years. Eskimo and Lapp cultures have lived as part of the ecosystem for thousands of years. Their numbers have been small, in keeping with the relatively low primary production of the tundra and adjacent seas. However, major disturbance to the permafrost terrain became evident in World War II, when military bases, airfields, and highways were hurriedly constructed without regard for maintenance of the natural protective surface. Recently, oil deposits were discovered on the north slope of Alaska with exploration and drilling creating possibilities for environmental damage. Oil spills on the landscape and in coastal waters are attendant problems.

Succession in the Arctic is a very slow process, and the development has the potential for disturbing the local system so that the ecosystem will not recover. The equilibrium of the tundra is based on very low energy flow. This means little chemical weathering, and slow rates of soil evolution and plant growth.

5.4   130 Subarctic Division

The source region for the continental polar air masses is south of the tundra zone between lat. 50° and 70°N. The climate type here shows great seasonal range in temperature. Winters are severe, and the region's small amounts of annual precipitation are concentrated in the three warm months. This cold, snowy, forest climate, referred to in this volume as the boreal subarctic type, is classified as *E* in the Köppen–Trewartha system. This climate is moist all year, with cool,

Fig. 5.6 Patterned ground caused by alternating freezing and thawing of the ground overlying permafrost, northeast of Fort Yukon, Alaska. Photograph by T.G. Freeman, Soil Conservation Service

short summers (see Fig. 5.2, climate diagram for Jakutsk, Siberia). Only 1 month of the year has an average temperature above 10 °C.

Winter is the dominant season of the boreal subarctic climate. Because average monthly temperatures are subfreezing for six to seven consecutive months, all moisture in the soil and subsoil freezes solidly to depths of a few meters. Summer warmth is insufficient to thaw more than a meter or so at the surface, so permafrost and patterned ground prevails over large areas (Fig. 5.6). Seasonal thaw penetrates from 0.5 to 4 m, depending on latitude, aspect, and kind of ground. Despite the low temperatures and long winters, the valleys of interior Alaska and Siberia were not glaciated during the Pleistocene, probably because of insufficient precipitation (Fig. 5.7).

The subarctic climate zone coincides with a great belt of needleleaf forest, often referred to as boreal forest, and open lichen woodland known as **tayga**. These species have adapted to the cold winter by greatly reducing their leaf area and by being able to respond rapidly to the short summer. Among the more widespread dominants are pine, fir, and spruce. Most trees are small, with more value to humans for pulpwood than for lumber. Different species occur in extremely wet and dry sites. In burned-over areas a mixture of deciduous trees and evergreen is characteristic

during secondary succession. In Siberia this second-growth enclave of broadleaf types is called "white tayga." Slow growing conifers reproduce the climax forest only over a long period of time.

The forests run diagonally across the continents. On the west coast, the boreal forest is more than 10° farther north than on the east coasts. The contrast of warm and cold ocean water on the two sides of the continents in higher middle latitudes causes this diagonal arrangement.

The Arctic needleleaf forest grows on **podzols** (Spodosols) with pockets of wet, organic Histosols. The podzol profile is distinctly shallower than any other mature profiles (see Fig. 6.1, p. 56), in a few places reaching depths greater then 45–60 cm. Soil development is slow because the land is frozen for long periods each year. For various reasons, notably the absence of earthworms, the humus layer on the surface is not mixed with the soil, but remains as a very black, highly acidic accumulation. The lower part of the A horizon is strongly leached to a gray or even white color. A distinct layer of humus and forest litter lies beneath the top soil layer. The B horizon is reddish from the accumulation of part of the leached material, and is very compact. Agriculture potential is poor, due to natural infertility of soils and the prevalence of swamps and lakes left by departed ice sheets. In some places, ice scoured the rock surfaces bare, entirely stripping off the overburden. Elsewhere rock basins were formed and stream courses dammed, creating countless lakes.

These lakes are only temporary features. Since decomposition is slow in the cold climate, these lakes gradually fill in with peat, organic matter produced by sphagnum moss or sedges, along with a definite succession of vegetation. These deposits have provided a low-grade fuel in northern Europe (Fig. 5.8).

Peat is also an excellent insulator. In the far north, it keeps summer heat from completely thawing the frozen ground below the depth of a 0.5 m or so. This subsoil, or permafrost, remains permanently frozen. Above it in the bog, annual freezing and thawing of the peaty soil pushes

Fig. 5.7 Extensive areas of Pleistocene glaciation are largely confined to the Northern Hemisphere. Adapted from *Glacial and Quaternary Geology* by R.F. Flint, p. 75. Copyright © 1971 by John Wiley & Sons, Inc. Reprinted by John Wiley & Sons, Inc.

Fig. 5.8 A peat bog in boreal forest near the border between Norway and Sweden. Photograph by John S. Shelton; from the University of Washington Libraries, Special Collections, John Shelton Collection, KC4567. Reproduced with permission

Fig. 5.9 Tilting of a line of poles in a bog is also caused by freezing and thawing, Yukon region, Alaska. The cross-poles at the base are to minimize the effect. Photograph by T.L. Pewe, U.S. Geological Survey

against the trees since it cannot push the solid permafrost down. This annual freeze-thaw cycle produces a topsy-turvy forest with tree trunks and utility poles leaning in all directions (Fig. 5.9).

The great north-flowing rivers of these regions, like those of the Arctic tundra, are subject to extensive spring floods, and the lowlands, even where they are not glaciated, do not dry out rapidly after the water recedes. Permanently water-logged surfaces, in many cases underlain by peat, are common in all these regions. In Canada, the name "muskeg" is applied to such surfaces.

This is the habitat of large ground animals—a population which derives most of its food supply from the aquatic life of the numerous lakes, rivers, and swamps. In these forests roam the world's chief fur-bearing animals: minks, martens, foxes, wolves, badgers, bears, beavers, squirrels, sables, and ermines. There are several large ungulata, chief of which are deer, moose, the caribou, and reindeer.

References

Brown J, Ferians OJ, Heginbotton JA, Melnikov ES (eds) (1997). Circum-arctic map of permafrost and ground ice conditions. U.S. Geological Survey, Scale 1:10.000,000, Washington, DC

Flint RF (1971) Glacial and quaternary geology. Wiley, New York, 892 pp

Köppen W (1931) Grundriss der Klimakunde. Walter de Gruyter, Berlin, 388 pp

Trewartha GT (1968) An introduction to climate, 4th edn. McGraw-Hill, New York, 408 pp

Walter H, Harnickell E, Mueller-Dombois D (1975) Climate-diagram maps of the individual continents and the ecological climatic regions of the earth. Springer, Berlin, 36 pp, with 9 maps

The Humid Temperate Ecoregions

6.1 200 Humid Temperate Domain

Both tropical and polar air masses govern the climate of the humid temperate domain, located in the mid-latitudes (30°–60°) on all the continents. The mid-latitudes are subject to **cyclones**; much of the precipitation in this belt comes from rising moist air along fronts within those cyclones. Pronounced seasons are the rule, with strong annual cycles of temperature and precipitation. The seasonal fluctuation of solar energy and temperature is greater than the diurnal (see Fig. 4.2, p. 29). The climates of the mid-latitudes have a distinctive winter season, which tropical climates do not. The lower temperatures of the winter season are due to two factors: reduced solar radiation and the inflow of cold air streams.

In the tropics (see Chap. 8), the different environments are distinguished on the basis of seasonal moisture pattern. There are regions with a high frequency of precipitation and regions with strong seasonal contrasts. Although the tropics are subject to one annual periodicity, most humid temperate regions are subject to two major annual cycles, one of solar energy and another of moisture.

In the temperate latitudes, probably the most important aspect of the hydrologic cycle is the periodic freezing of lakes, streams, and soil moisture. This freezing stops the flow of water runoff. At the same time, it decreases the water supply available to the plants, producing drought.

Most streams will reflect the winter season in the flow pattern. When the spring thaw occurs, flooding may occur if there are large amounts of snow in the watershed and melting takes place rapidly. Removal of headwater forests leads to increased flooding.

Many of the same process of soil development which operate in tropical forest lands are active here, but go on more slowly, owing to lower temperatures and less extreme humidity. Humus accumulates, too, for the slower decay of organic litter on the forest floor, when mixed with the soil layers, imparts a brownish color.

Three zonal soil types are recognized in the regions of this group (Fig. 6.1). The red and yellow lateritic soils of the humid tropical group extend poleward into the warmer parts of the humid temperate lands, and are known as **yellow forest soils** (Ultisols). Farther poleward, however, humus accumulation is sufficiently rapid so that soil color is darkened. With the aid of earthworms, the organic matter is mixed with the upper soil layers to form the **brown forest soils** (Alfisols). The podzols (Spodosols) lie on the northern borders of this group, extending into the polar group. In the profiles of the podzol, the absence of earthworms is indicated by the concentration of humus at the surface, and the light, ashy color of the soil below (the name podzol is derived from a Russian word meaning ashes). The depth of these profiles decreases as the length of the frozen period of winter increases. None of these soils are fertile.

R.G. Bailey, *Ecoregions*, DOI 10.1007/978-1-4939-0524-9_6, © Springer Science+Media, LLC 2014

Fig. 6.1 Generalized
mature soil profiles
developed under mid-
latitude forests. After Jenny
(1941) in James (1959)
p. 197

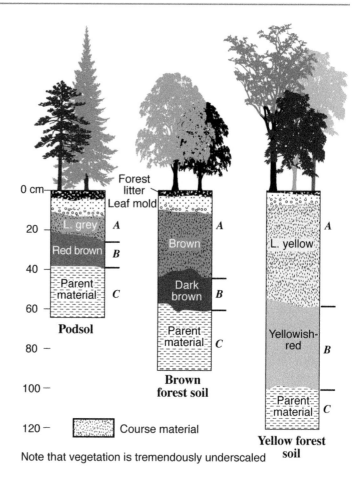

Note that vegetation is tremendously underscaled

The regions of the humid temperate domain occur within climatic conditions where there is a winter cold season when plant growth ceases, and rain in summer is sufficient to support forest vegetation of broadleaf deciduous and needleleaf evergreen trees. On the equatorward side of these regions where this group borders the tropical regions, the differences in the forest are not sharply contrasted. Gradually the species that cannot survive frost drop out. On the poleward side of these regions in the Northern Hemisphere, there is a relatively sharp line of demarcation. The boreal forest occurs where there are severe winters and only short, cool summers. Here the spruce, fir, and larch are the more widespread conifers. Toward the continental interiors, on the dry side of these regions, grasslands usually border the forests.

Animals in these climates need to survive cold winters and seasonal variations in their food supply. Some vary their diet. Many birds and mammals migrate to warmer climates. Most amphibians and reptiles, as well as some large mammals, hibernate or become dormant.

The variable importance of winter frost determines six divisions: *warm continental, hot continental, subtropical, marine, prairie,* and *mediterranean.* Figure 6.2 shows the distribution of these divisions. Climate diagrams for these divisions are presented in Fig. 6.3.

6.2   210 Warm Continental Division

South of the eastern area of the subarctic climate, between latitudes 40° and 55°N and from the

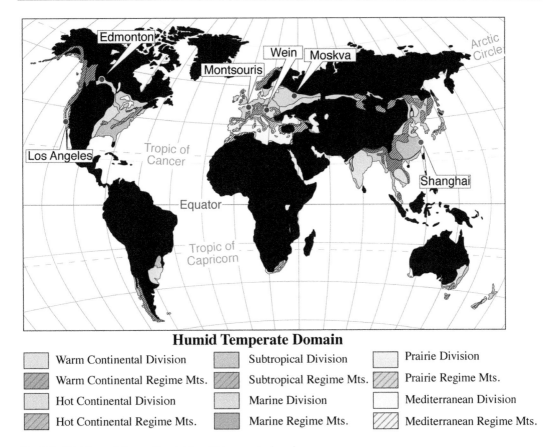

Humid Temperate Domain

Warm Continental Division	Subtropical Division	Prairie Division
Warm Continental Regime Mts.	Subtropical Regime Mts.	Prairie Regime Mts.
Hot Continental Division	Marine Division	Mediterranean Division
Hot Continental Regime Mts.	Marine Regime Mts.	Mediterranean Regime Mts.

Fig. 6.2 Divisions of the continental humid temperate domain

continental interior to the east coast, lies the humid, warm-summer, continental climate. Located squarely between the source regions of polar continental air masses to the north, and maritime or continental tropical air masses to the south, it is subject to strong seasonal contrasts in temperature as air masses push back and forth across the continent.

This climate occurs only in the Northern Hemisphere. It applies to the northeastern United States and southeastern Canada, southeastern Siberia, and northern Japan. Very similar climatic conditions also hold for eastern Europe across the Baltic countries and Russia as far as the Urals.

The Köppen-Trewartha system designates this area as *Dcb*, a cold, snowy, winter climate with a warm summer (see Fig. 6.3, climate diagram for Moskva, Russia). This climate has 4–7 months when temperatures exceed 10 °C, with no dry season. The average temperature during the coldest month is below 0 °C. The warm summer signified by the letter *b* has an average temperature during its warmest month that never exceeds 22 °C. Precipitation is ample all year, but substantially greater during the summer. In eastern Asia, a monsoon effect is strongly accentuated in summer.

Mixed boreal and deciduous forest grows throughout the colder northern parts of the humid continental climate zone (Fig. 6.4), and is therefore transitional between the boreal forest to the north and the deciduous forest to the south. In eastern North America, part of it consists of mixed stands of a few coniferous species (mainly pine) and few deciduous species (mainly birch, maple, and beech). The rest is a macromosaic of pure deciduous forest in favorable habitats, with good soil and pure coniferous forest in less favorable habitats with poor soils. Here soils are

Fig. 6.3 Climate diagrams from the various divisions of the humid temperate domain: mixed deciduous–coniferous forest regions (continental with cold winters and warm summers) and deciduous forest region (more moderate), broadleaf evergreen forest region (very rainy with hot summers), oceanic broadleaf forest region (rainy with warm summers), prairie region (cold winter with warm summer), and sclerophyllous regions of California (dry summer). Redrawn from Walter et al. (1975)

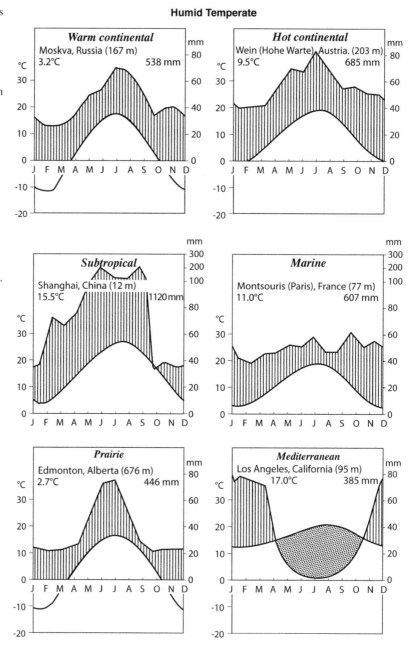

podzols (Spodosols). Such soils have a low supply of bases and a horizon in which organic matter and iron and aluminum have accumulated. They are strongly leached, but have an upper layer of humus. Cool temperatures inhibit bacterial activity that would destroy this organic matter in tropical regions. Deficient in calcium, potassium, and magnesium, soils are generally acidic. Thus, they are poorly suited to crop production, even though adequate rainfall is generally assured. Conifers thrive here.

Because of the availability of soil water through a warm-summer growing season, this environment has an enormous potential for food production. North America and Europe support dairy farming on a large scale. A combination of

Fig. 6.4 Broadleaf deciduous and needleleaf evergreen forest as it appears in the Lake States region of north central United States. Photograph by Robert G. Bailey

Fig. 6.5 A stand of mature sugar maples in the Allegheny National Forest, Pennsylvania. Photograph by B.W. Muir, U.S. Forest Service

acid soils and unfavorable glacial terrain in the form of bogs and lakes, rocky hills, and stony soils has deterred crop farming in many parts.

6.3   220 Hot Continental Division

South of the warm continental climate lies another division in the humid tropical domain, one with a humid, hot-summer continental climate. It has the same characteristics as the warm continental except that it is more moderate and has hot summers and cool winters (see Fig. 6.3, climate diagram for Wien, Austria). The boundary between the two is the isotherm of 22 °C for the warmest month. In the warmer sections of the humid temperate domain, the frost-free or growing season continues for 5–6 months, and in the colder sections only 3–5 months. Snow cover is deeper and lasts longer in the northerly areas.

This climate is located in central and eastern parts of United States, northern China (including Manchuria), Korea, northern Japan, and central and eastern Europe.

In the Köppen-Trewartha system, areas in this division are classified as *Dca* (*a* signifies hot summer). We include in the hot continental division the northern part of Köppen's *Cf* (subtropical) climate region in the eastern United States. Köppen used as the boundary between the *C-D* climates the isotherm of −3 °C for the coldest month. Thus, for example, in the United States,

Köppen places New Haven, Connecticut, and Cleveland, Ohio, in the same climatic region as New Orleans, Louisiana, and Tampa, Florida, despite obvious contrasts in January mean temperatures, soil groups, and natural vegetation between these northern and southern zones. Trewartha (1968) redefined the boundary between *C* and *D* climates as the isotherm of 0 °C of the coldest month, thereby pushing the climate boundary south to a line extending roughly from St. Louis to New York City. Trewartha's boundary is adopted here to distinguish between humid continental and humid subtropical climates.

Natural vegetation in this climate is winter deciduous forest, dominated by tall broadleaf trees that provide a continuous dense canopy in summer, but shed their leaves completely in the winter (Fig. 6.5). Lower layers of small trees and shrubs are weakly developed. In spring, a luxuriant ground cover of herbs quickly develops, but is greatly reduced after trees reach full foliage and shade the ground. Common trees in eastern United States, eastern Europe, and eastern Asia are oak, beech, birch, hickory, walnut, maple, basswood, elm, ash, tulip, sweet chestnut, and hornbeam. Hemlock, a needleleaf evergreen tree, may also be present.

Soils are chiefly **red-yellow podzols** (Ultisols), and **gray-brown podzols** (Alfisols), rich in humus and moderately leached with a distinct, light-colored zone under the upper dark layer. The red-yellow podzols (Ultisols) have a low supply of bases and a horizon of accumulated clay. Where topography is favorable, diversified farming and dairying are the most successful agricultural practice. Because the gray-brown podzols (Alfisols) are soils of high base status, they proved highly productive for farming after the forests were cleared.

Most of the native forest has been cleared. The forest is preserved throughout the mountainous terrain of the Appalachians Mountains and the woodlots throughout the farmed belt. Much of this forest consists of second- or third-growth tree stands. Many farms in the forested area were abandoned and have since been covered by successional forest. The original animal life was abundant with deer, bear, panthers, squirrels, and wild turkeys. Large mammals became scarce during the peak of agricultural use, but populations have increased with return of the forest environment.

In China and Korea, the effects of prolonged deforestation are evident everywhere. Throughout central Europe, large areas have been under field crops and pastures for centuries, while at the same time forests have been cultivated. Cereals grown extensively in North America and Europe include corn and wheat. In north China, wheat is the principal crop. Rice is the dominant crop in both south Korea and Japan. Soybeans are intensively cultivated in the midwestern United States and in northern China and Manchuria, but very little in Europe.

6.4   230 Subtropical Division

The humid subtropical climate, marked by high humidity (especially in summer) and the absence of really cold winters, prevails on the eastern sides of the five continents in lower middle latitudes, and is influenced by trade and monsoon winds. These areas include the southeastern United States, southern China, Taiwan (Formosa), southernmost Japan, Uruguay and

adjoining parts of Brazil and Argentina, the eastern coast of Australia, and the North Island of New Zealand.

In the Köppen-Trewartha system, this area lies within the *Cf* climate, described as temperate and rainy with hot summers (see Fig. 6.3, climate diagram for Shanghai, China). The *Cf* has no dry season; even the driest summer month receives at least 30 mm of rain. The average temperature of the warmest month is warmer than 22 °C. Rainfall is ample all year, but is markedly greater during summer. Rivers and streams flow copiously through much of the year. Thunderstorms, whether of thermal, squall-line, or cold-front origin, are especially frequent in summer. Tropical cyclones and hurricanes strike the coastal area occasionally, always bringing heavy rains and flooding. Winter fronts bring precipitation, some in the form of snow. Temperatures are moderately wide in range, comparable to those in tropical deserts, but without the extreme heat of a desert summer.

Soils of the moister, warmer parts of the humid subtropical regions are strongly leached red-yellow podzols (Ultisols) related to those of the humid tropical and equatorial climates. Rich in oxides of both iron and aluminum, these soils are poor in many of the plant nutrients essential for successful agricultural production. They are susceptible to severe erosion and gullying when exposed to forest removal and intensive cultivation.

Forest is the natural vegetation throughout most areas of this division. Along the outer coastal plain of the United States and in a large part of southern China and the south island of Japan, the native broadleaf forest was of the evergreen type. This forest consists of trees such as the evergreen oak, and trees of the laurel and magnolia families. Near its northern limits, vegetation of this region grades into broadleaf deciduous forest. Another type of rainforest is found in southeastern Australia and Tasmania and consists of many species of eucalyptus, which may reach heights of 100 m (Fig. 6.6). The rainforest flora found in New Zealand consists of large tree ferns, large conifers such

Fig. 6.6 Eucalyptus forest in southeastern Australia. Photograph by Forests Commission of Victoria; from the American Geographical Society Library, University of Wisconsin-Milwaukee Libraries

as the kauri tree, podocarp trees, and small-leaved southern beeches.

Broadleaf evergreen forest may have a well-developed lower layer of vegetation that may include tree ferns, small palms, bamboos, shrubs, and herbaceous plants. Lianas and epiphytes are abundant.

Much of the southeastern United States today is covered by a second-growth forest consisting of a number of pine species. It grows on the sandy soils of the coastal plain, where it appears to be a specialized type dependent on fast-draining sandy soils and frequent fires for its preservation.

Today, large areas have been converted to agricultural croplands, particularly in China. Corn is a major crop in the southern United States. Cattle production and tree farming are the important uses of soils too sandy for field crops. Rice and tea are the most important crops in those parts of China and Japan with similar climate.

6.5   240 Marine Division

Situated chiefly on the continental west coasts and on islands of the higher middle latitudes between 40° and 60°N is a zone that receives abundant rainfall from maritime polar air masses, and has a narrow range of temperature because it borders on the ocean. These coasts and islands are bathed by warm ocean water, and the prevailing westerly winds bring abundant moisture to the land.

Trewartha (1968) classified the marine, west coast climate as *Do*—temperate and rainy, with warm summers. The average temperature of the warmest month is below 22 °C, but at least 4 months of the year have an average temperature of 10 °C. The average temperature during the coldest month of the year is above 0 °C. Precipitation is abundant throughout the year, but is markedly reduced during the summer (see Fig. 6.3, climate diagram for Montsouris [Paris], France). Although total rainfall is not great by tropical standards, the cooler air temperatures reduce evaporation and produce a damp, humid climate with much cloud cover. Mild winters and relatively cool summers are typical. Coastal mountain ranges influence precipitation markedly in these middle latitudes. The mountainous coasts of British Columbia and Alaska annually receive 1,530–2,040 mm of precipitation and more. Heavy precipitation greatly contributed to the development of fiords along the coast in Norway (Fig. 6.7), British Columbia, southern Chile, and the South Island of New Zealand. Heavy snows in the glacial period fed vigorous valley glaciers that descended to the seas, scouring deep troughs that reach below sea level at their lower ends. Farther back from the coast the annual rainfall decreases, even to less than 75 cm (London 60 cm; Paris 55 cm); but in the absence of any very high temperatures so little evaporation occurs that even this relatively small amount is highly effective and supports a luxuriant plant growth.

Needleleaf forest is the natural vegetation of the marine division. In the coastal ranges of the

Fig. 6.7 This Norwegian fiord has the steep rock walls of a deep glacial trough. Postcard by Eneret Mittet and Co. [Place of publication unknown, author's collection]

northwestern United States, it is the redwood zone. Farther north, this vegetation is succeeded by Douglas-fir, western red cedar (Fig. 6.8), and spruce which grow to enormous heights, forming some of the densest of all coniferous forest with some of the world's largest trees. Under the lower precipitation regime of Ireland, southern England, France, and the Low Countries, a broadleaf deciduous forest was the native vegetation. However, much of it disappeared many centuries ago under cultivation, so that only scattered forest plots or groves remain. Dominant tree species of this forest type in western Europe are oak and ash, with beech in the cooler, moister areas. In the Southern Hemisphere, the forests of Tasmania and New Zealand, and the mountainous coastal belt of southern Chile are of the **temperate rainforest** class.

The trees are covered with mosses, epiphytes, and ferns. They cover everything from living branches and leaves to rotting logs. Because of the cool climate, decomposition works slowly; log piles upon logs to form a moss and fern-covered jumble over the forest floor, which is deep in humus. This is very much unlike the tropical rainforest, where decomposers make short work of falling debris so that the forest floor has little humus and few fallen logs.

Because of recent Pleistocene glaciation, landforms of glacial erosion and deposition are little changed from their original shapes. Glacial troughs and fiords are striking landforms. Extensive lowlands of northwestern Europe consist of till plains, moraines, and outwash plains left by ice sheets. Mountain watersheds, when disturbed

Fig. 6.8 Damp, oceanic coniferous forest with western red cedar in Mt. Baker National Forest, Washington. Photograph by E. Lindsay, U.S. Forest Service

by logging, can experience severe erosion and high sediment yields, particularly from mass wasting.

Soils of the marine regions bearing needleleaf forest are strongly leached, acidic brown forest soils (Alfisols). Due to the region's cool temperatures, bacterial activity is slower than in the warm tropics, so unconsumed vegetative matter forms a heavy surface deposit. Organic acids from decomposing vegetation react with soil compounds, removing bases such as calcium, sodium, and potassium. Under the lower precipitation of the British Isles and western Europe, mid-latitude deciduous forests are underlain by gray-brown podzols (Alfisols).

Productivity is fairly high in these forests, but it chiefly goes into wood. Available forage is not high, so that the biomass of large animals is not great, although deer, elk, and mountain lions form part of the American marine ecosystem. In New Zealand, no mammals or reptiles existed in the original forest. The niches were filled by birds, such as the kiwi.

The marine regions of western Europe and the British Isles have been intensively developed for centuries for such diverse use as crop farming, dairying, orchards, and forests. In North America, the mountainous terrain is not suitable for agricultural use, except in limited valley floors. Forests are the primary resource here and constitute the greatest structural and pulpwood timber resource on Earth. Douglas-fir, western cedar, and western hemlock are the principal lumber trees. The same mountainous terrain that limits agriculture is a producer of enormous water surpluses that run to the sea in rivers. These rivers support anadromous fisheries.

In spite of the heavy rainfall, periods of drought up to several weeks can occur, so that many areas in the northwestern United States and the western coast of New Zealand have been burned. In New Zealand, the introduction of many species of deer proved to be a serious mistake, since these forests evolved in the absence of grazing mammals. Deer and sheep have seriously overbrowsed these forests.

6.6   250 Prairie Division

The deciduous forests of the temperate zone are confined to climatic regions of an oceanic nature, where extremes of temperature are not sharp, and where rainfall is more or less evenly distributed throughout the year. Two chief kinds of grasslands thrive in the transition zone of the middle latitudes between the forests and the

Fig. 6.9 Prairie parkland on the Central Lowland, in northwest Iowa. Photograph by Robert G. Bailey

deserts. On the dry margins are the short-grass **steppes** (p. 75) and on the wet margins are the tall-grass **prairies**.

Prairies are typically associated with continental, mid-latitude climates designated as *sub-humid*. Precipitation in these climates ranges from 510 and 1,020 mm per year, and is almost entirely offset by evapotranspiration (see Fig. 6.3, climate diagram for Edmonton, Alberta). In summer, air and soil temperatures are high. Soil moisture in the uplands is inadequate for tree growth, and deeper sources of water are beyond the reach of tree roots. In North America, prairies form a broad belt extending from Texas northward to southern Alberta and Saskatchewan. Similar prairies occur in the humid *pampa* of Argentina, in Uruguay and southern Brazil, in the *puszta* of Hungary and on the northern side of Russian steppes, in South Africa, in Manchuria, and in Australia.

In a transitional belt on the wetter border of the division, forest and prairie mix in the so-called **forest-steppe** or **parkland**. It is not a homogeneous vegetation formation like the tropical savanna (Chap. 8), but rather a macromosaic of deciduous forest stands and prairie. Relief and soil texture determine the predominating vegetation. Forests are found on well-drained habitats, slightly raised ground, the sides of the river valleys, and porous soils, whereas the prairies occupy badly drained, flat sites with a relatively heavy soil (Fig. 6.9). In the western prairies of the United States, the grassland has been changed with the introduction of trees. Forests were planted around the farmsteads and villages, so that today the buildings are all but hidden in foliage during the summer.

In Canada and northern United States the transition from grassland to boreal forest consists of a narrow belt of poplar and aspen deciduous forest. This belt is from 60 to 200 km wide.

The boundary between the prairie and the forest is not so clearly related to climatic or edaphic conditions. Prairies exist well within a climate humid enough for tree growth, and where trees are planted they grow if protected from competition with the roots of the grasses. In previous times, fires caused by lightning and the grazing of big-game herds encouraged the growth of the grasses in the treeless, wet prairies.

The prairie climate is not designated as a separate variety in the Köppen-Trewartha system. Geographers' recognition of the prairie climate (Thornthwaite 1931; Borchert 1950) has been incorporated into the system presented here. Prairies lie on the arid western side of the humid continental climate, extending into the subtropical climate at lower latitudes. Temperature characteristics correspond to those of the adjacent humid climates, forming the basis for two types of prairies: temperate and subtropical.

Tall grasses associated with subdominant broad-leaved herbs dominate prairie vegetation. The grasses of the Argentine Pampa are said to have once risen above the head of a man on horseback. Trees and shrubs are almost totally absent, but a few may grow as woodland patches in valleys and other depressions. Deeply rooted grasses form a continuous cover. They flower in spring and early summer, the forbs in late summer. In the tall-grass prairie of Iowa, for example, typical grasses are big bluestem and little bluestem; a typical forb is black-eyed Susan.

Because rain falls less in the grasslands than in forest, less leaching of the soil occurs. The pedogenic process associated with prairie vegetation is **calcification**, as carbonates accumulate in the lower layers. Soils of the prairies are prairie soils (Mollisols), which have black, friable, organic surface horizons and a high content of bases. Grass roots deeply penetrate these soils.

Bases brought to the surface by plant growth are released on the surface and restored to the soil, perpetuating fertility. These soils are the most productive of the great soil groups.

These soils are not uniform and reflect the transitional nature of the climate. A succession of soil types, from the humid forest margins across the prairies and the steppes to the dry lands, conforms to the changes in moisture and vegetation cover (Fig. 6.10). On the rainy margins of the prairie, a deep soil is formed which is so abundantly supplied with organic material that is dark-colored even in the B horizon. This is the **black prairie soil** (Fig. 6.11). Near the dry margin, rainfall decreases to the point that minerals dissolved near the surface are carried down to the B horizon and no farther.

Two soil types share this process of having mineral accumulations, chiefly lime, in the B horizon. The first of these, occupying the dry margins of the prairies, is known as the **chernozem** (Mollisol). The color of the chernozem is even darker than the black prairie soil, and its fertility is increased by the decreased effectiveness of the leaching process. The dry boundary of the chernozem coincides with the prairie–steppe boundary, where, because the depth of the moist surface soil becomes less than about 60 cm, the tall grasses give way to the short grasses. The smaller supply of humus from the short grasses is reflected in a change from the black color of the chernozem to a chestnut-brown color; and the more active evaporation and shallower penetration of the rain water result in the formation of a continuous layer of lime salts much closer to the surface than in the chernozem. This is the **chestnut-brown soil** (Mollisol).

Most of the regions of this type are either plains or plateaus. The rainfall in the prairie is usually sufficient to support permanent streams, many of which are lined by **galeria forests**.

Most of the prairie has disappeared, replaced by some of the richest farmland in the world (Fig. 6.12). Some of the native animals of the region enjoyed either an elimination of their natural enemies or an increase in the supply of food, which made possible a sudden and large increase in their numbers.

Fig. 6.10 Generalized mature soil profiles that develop under mid-latitude grassland. From Jenny (1941) in James (1959) p. 312

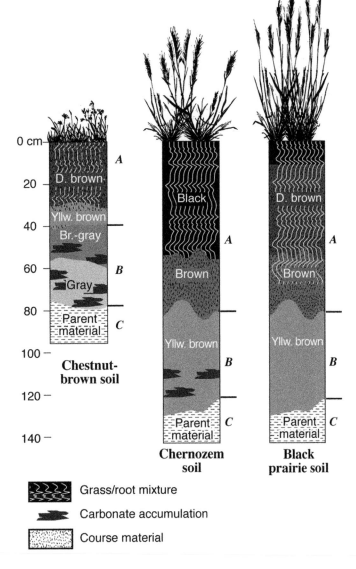

Chestnut-brown soil

Chernozem soil

Black prairie soil

Grass/root mixture

Carbonate accumulation

Course material

Fig. 6.11 Corn planted in prairie soil, Iowa. Photograph by Robert G. Bailey

The agricultural region that developed on the central plains of the United States as a result of the spread of farm settlement on to the prairie is known as the Corn Belt. Here the practice of feeding maize to hogs and cattle, and marketing the fattened animals is sustained by the vast areas of level land, and the remarkably sustained fertility of the soils. Although maize is the principal crop worldwide, this region is also ideally suited to other crops, such as wheat, particularly in Russia. On the other hand, soybeans are intensively cultivated in midwestern United States and in northern China and Manchuria.

Fig. 6.12 North American Corn Belt in a prairie region of South Dakota. Some patches of relict galeria forest follow the stream in the foreground. Photograph by John S. Shelton; from the University of Washington Libraries, Special Collections, John Shelton Collection, KC6162

6.7   260 Mediterranean Division

Situated on the western margins of the continents between latitudes 30° and 45°N is a zone subject to alternately wet and dry seasons, the transition zone between the dry west coast desert and the wet west coast. There are five such locations in the world. The largest borders the Mediterranean Sea. In North America the area included in this division lies primarily in California. Other areas are found in Chile, in South Africa around Capetown, and in Australia, where they are divided into a western area around Perth and an eastern area around Adelaide. They occupy only 2 % of the Earth's surface.

Trewartha (1968) classified the climate of these lands as *Cs*, signifying a temperate, rainy climate with dry, hot summers. The symbol *s* signifies a dry summer (see Fig. 6.3, climate diagram for Los Angeles, Calif.).

This climate is a product of subsidence associated with the subtropical high. In the summer the high moves poleward over these areas bringing essentially desert weather. In the winter the anticyclonic circulation moves equatorward allowing the westerlies to bring moisture into the area. These weather patterns often lead to summer wildfires (Fig. 6.13).

Fig. 6.13 Fire sweeping through California chaparral in summer. Photograph by Leonard F. DeBano, U.S. Forest Service

The combination of wet winters with dry summers is unique among climate types, and produces a distinctive natural vegetation of hard-leaved evergreen trees and shrubs called **sclerophyll**, scrub woodland. They reduce water loss with leaves which are small, thick, and stiff, with hard, leathery, and shiny surfaces. Various forms of sclerophyll woodland and scrub are also typical. Trees and shrubs must withstand the severe summer drought—two to four rainless months—and severe evaporation. Although in the different continents different species compose the woodland, the appearance of the

Fig. 6.14 Sclerophyll open woodland south of San Francisco, California. Most of the trees are oaks. Photograph by R.E. Wallace, U.S. Geological Survey

vegetation, resulting from its adaptation to the peculiarities of climate, is strikingly similar. The broadleaf evergreen woodlands of southern Europe, for example, are composed mostly of various kinds of oaks, whereas the similar forests of Australia are species of eucalpytus.

Because the winters are not cold enough, nor the summer droughts long enough to enforce a period of rest, there is no season when the leaves drop from the trees and growth ceases. In this way the mediterranean vegetation differs from selva (p. 84), which is evergreen but which has no seasonal rhythm. The woodland adapts itself to summer droughts. The trees are widely spaced and all plants have deep tap roots and a wide development of surface roots (Fig. 6.14). The evaporation from the plants is diminished by a thick bark and by the sclerophyllous leaves.

In many areas, the original cover of woodland has been radically altered, probably by humans. At present, large areas of this group are covered by a thick, low growth of bushes and shrubs, known as *maquis* in Europe, *chaparral* in California, and *mallee* in southern Australia. Extreme flammability characterizes the chaparral during the long, dry summer. This poses an ever

present threat to suburban housing which has expanded into chaparral-covered hillsides in California.

This has raised a problem of resource management in the suburban areas around cities. When an attempt is made to keep fires from starting, the brush grows thicker and accumulates a layer of debris underneath. As a result, the brush fires are more destructive than they were when smaller fires occurred each year. Furthermore, when a large fire has completely burned the cover, the torrential winter rains that follow produce mud flows and floods. The soil may be swept completely away, leaving only the bare rock exposed at the surface, while bordering valleys are filled with mud. This situation is worsened in southern California where soil becomes water repellent following fire.

After a fire, many mediterranean shrubs resprout from root crowns. The seeds of some species need fire to germinate and may lay dormant for years until the next fire.

Soils of this mediterranean climate are not susceptible to simple classification. Soils typical of semiarid climates associated with grasslands are generally found. Severe and prolonged soil erosion following deforestation and overgrazing has left the mediterranean region with much exposed regolith and bedrock.

Animals survive fire by taking flight, or by retreating to underground burrows. Native mammals include deer, rabbits, and numerous rodents. In southern Europe and California, many of the native species have been replaced by large domestic grazers such as cows, sheep, and goats. Much of the birdlife is migratory, visiting mainly during the spring and fall. Resident birds tend to have short wings and long tails, an aid to maneuvering around shrubs.

The region is an important source of citrus fruits, grapes, and olives. In the mediterranean, cork from the bark of the cork oak is also important. In central and southern California citrus, grapes, avocadoes, nuts (almond, walnut), and deciduous fruits are extensively grown. Irrigated alluvial soils are also highly productive for vegetable crops such as carrots, lettuce, artichokes, strawberries, and forage crops (alfalfa).

This division lies closely hemmed in between high mountains and the sea. The surface features include small and isolated valley lowlands, bordered by hills and backed by high mountain ranges. The heavy rains in the highlands feed numerous torrential streams. The gravels and sands which they bring down with them to the lowlands are piled up in huge alluvial fans along the piedmonts. Delta plains grow where the rivers flow into the sea. Owing to the concentration of rain in the winter season, the regimen of these streams shows a maximum in that season. However, where the streams rise high enough in the mountains to reach the snow fields, the maximum flow comes during the melting period in the spring. The removal of forest from the mountains has seriously changed this regimen in many areas. The removal of the forest causes severe floods during the winter and spring, and during the summer these floods are followed by droughts, when the streams dry up. The original forest cover of most of the regions bordering the Mediterranean Sea has largely been removed, seriously affecting the habitability of the lowlands.

References

Borchert JF (1950) The climate of the central North American grassland. Ann Assoc Am Geogr 40:1–39

James PE (1959) A geography of man, 2nd edn. Ginn, Boston, 656 pp

Jenny H (1941) Factors of soil formation. McGraw-Hill, New York, 281 pp

Thornthwaite CW (1931) The climates of North America according to a new classification. Geogr Rev 21:633–655, with separate map at 1:20,000,000

Trewartha GT (1968) An introduction to climate, 4th edn. McGraw-Hill, New York, 408 pp

Walter H, Harnickell E, Mueller-Dombois D (1975) Climate-diagram maps of the individual continents and the ecological climatic regions of the earth. Springer-Verlag, Berlin, 36 pp, with 9 maps

The Dry Ecoregions

7.1   300 Dry Domain

The essential feature of a dry climate is that annual losses of water through evaporation at the Earth's surface exceed annual water gains from precipitation. Due to the resulting water deficiency, no permanent streams originate in dry-climate zones. Because evaporation, which depends chiefly on temperature, varies greatly from one part of the Earth to another, no specific value for precipitation can be used as the boundary for all dry climates. For example, 610 mm of annual precipitation produces a humid climate and forest cover in cool northwestern Europe, but the same amount in the hot tropics produces semiarid conditions.

The tropical dry climates occupy the air masses in the subtropical high-pressure cells centered over the tropics of Cancer and Capricorn, both north and south of the equator, in the zone between 20° and 30°. This subsiding air mass is stable and dry. Dry land regions can be found on the western sides of all the continents. The dry lands also extend inland from the western sides of the continents, bending poleward in each hemisphere. The general regularity of the global pattern of dry climates is a reflection of the regular arrangement of certain climatic and water features. Fig. 4.6 (p. 31) shows that cold water baths parts of the west coasts of all the continents. Because evaporation is much less from cold water than warm water, the rainfall is much less here than continental margins bathed by warm water.

The presence of dry lands along the east coast of South America in Patagonia is associated with a wide expanse of cold water of the South Atlantic Ocean (Falkland Island Current). In North America the dry lands cannot extend so far toward the east as they do in Asia because of the movement of moist maritime air from the Gulf of Mexico up to the Mississippi Valley, and the lack of such air moving across the Himalayas, in Asia.

We commonly recognize two divisions of dry climates: the arid *desert* (*BW*) and the semiarid *steppe* (*BS*). Generally, the steppe is a transitional belt surrounding the desert and separating it from the humid climates beyond. The boundary between arid and semiarid climates is arbitrary, but commonly defined as one-half of the amount of precipitation separating steppe from humid climates. These climates are displayed in Fig. 7.1 and mapped in Fig. 7.2.

Of all the climatic groups, dry climates are the most extensive; they occupy a fourth or more of the Earth's land surface (see Fig. 4.7, p. 32).

In these climates, many plants and animals have adapted to live with minimal rain, drying winds, and high temperatures.

7.2   310 Tropical/Subtropical Steppe Division

Tropical steppes occur along the less arid margins of the tropical deserts on both the north

R.G. Bailey, *Ecoregions*, DOI 10.1007/978-1-4939-0524-9_7, © Springer Science+Media, LLC 2014

Fig 7.1 Climate diagrams of steppe and desert stations: *Above*: central Asia, with some rain at all seasons, and winter rain. *Below*: with summer rain (northern Africa), and with rain that may fall at any season (Atacama Desert). Redrawn from Walter et al. (1975)

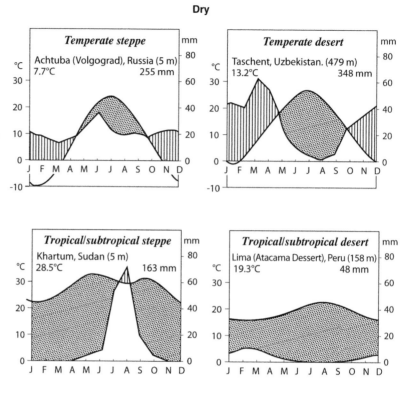

and south, and in places on the east as well. On the equatorward side of the deserts it is transitional to the wet-dry tropical climate. Locally, altitude causes a semiarid steppe climate on plateaus and high plains that would otherwise be desert. Steppes on the poleward fringes of the tropical deserts grade into the Mediterranean climate in many places. In the United States, they are cut off from the Mediterranean climate by coastal mountains which allow the tropical deserts to extend farther north. Other important steppes of this type are the interior of the Kalahari Desert of South Africa, the dry eastern piedmont of the Andes, and northeastern Brazil.

Trewartha (1968) classified the climate of tropical/subtropical steppes as *BSh*, indicating a hot, semiarid climate where potential evaporation exceeds precipitation, and where all months have temperatures above 0 °C (see Fig. 7.1, climate diagram for Khartum, Sudan). Average rainfall is from 25 to 76 cm annually.

Steppes typically are grassland of short grasses and other herbs, with locally developed shrub and woodland. In the United States, pinyon-juniper woodland grows on the Colorado Plateau, for example. To the east, in New Mexico and Texas, the grasslands grade into savannah woodland or semi-deserts composed of xerophytic shrub and trees, and the climate becomes nearly arid-subtropical. Cactus plants are present in some places.

In the tropics, semi-desert is associated with this climate. A particularly important occurrence is the thorntree savanna of Africa, characterized by thorny trees and shrubs that shed their leaves for the long dry season. Another important area of this type is found in the Kalahari Desert of southern Africa, the home of the Bushmen. In the semi-desert zone of the African Sahel, the climate is associated with the acacia-desert grass savannah where the stunted trees stand far apart and the short desert grasses cover most of the

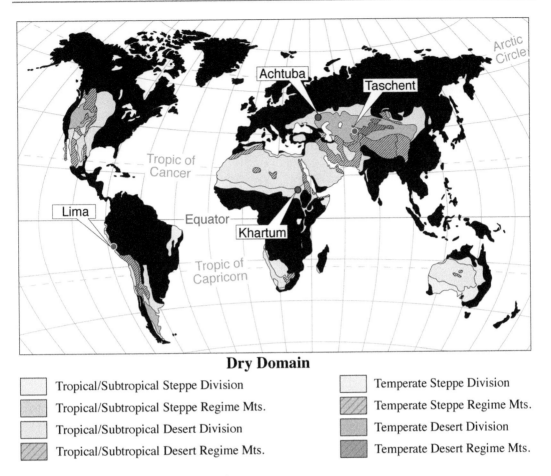

Dry Domain

Tropical/Subtropical Steppe Division

Tropical/Subtropical Steppe Regime Mts.

Tropical/Subtropical Desert Division

Tropical/Subtropical Desert Regime Mts.

Temperate Steppe Division

Temperate Steppe Regime Mts.

Temperate Desert Division

Temperate Desert Regime Mts.

Fig 7.2 Divisions of the continental dry domain

surface. In northwestern Brazil, there is an area once covered with thorny, scrubby deciduous trees known as *caatinga*. Where brush is thicker and mixed with trees it forms the scrub woodland of the Chaco in Argentina. Scrub woodland also forms a fringe around the desert in Australia (Fig. 7.3).

Chestnut-brown soils and sierozems (Mollisols, Aridisols) are associated with these tropical, semi-arid climates.

In Africa, nomadic herding is one of the main forms of agrarian uses, which is linked to the alternation of the dry and rainy seasons. During the dry season the herders move into high-altitude regions, which are generally wetter, and during the rainy season they move down to the lowlands again. In the deserts and semi-deserts,

the herds consist of camels, sheep, and goats, while in the thorn savannas, cattle predominate.

Like the temperate steppe we describe below, rainfall in these regions can be expected to vary greatly and is subject to periods of drought interspersed with periods of ample rainfall. In several West African nations within the Sahelian zone, recent droughts have depleted grasses for grazing and devastated the annual grain crop. Some five million cattle perished and many thousands of people died of starvation and disease. Periodic droughts in the past are well documented. In places, the land surface has been changed into desert through a process called **desertification**. Desertification in the African steppes can be attributed to greatly increased numbers of humans and their cattle.

Fig 7.3 Mulga scrub, a subtropical semi-desert, near Oudabunna Sta., western Australia. Photograph by R.A. Gould. Image #335754, American Museum of Natural History, Library

Fig 7.4 The incised channel of the Rio Puerco, New Mexico, northwest of Albuquerque. It carries water only a few days each year. Photograph by Lev Ropes, Guru Graphics

7.3   320 Tropical/Subtropical Desert Division

The continental desert climates are south of the Arizona-New Mexico Mountains in the United States. They are not only extremely arid but also have extremely high air and soil temperatures. Direct solar radiation is very high, as is outgoing radiation at night, causing extreme variations between day and night temperatures, and a rare nocturnal frost. Annual precipitation is less than 200 mm, and less than 100 mm in extreme deserts (see Fig. 7.1, climate diagram for Lima, Peru). These areas have climates that Trewartha (1968) calls *BWh*.

The vast desert belt extending across North Africa (Sahara Desert), Arabia, and Iran to Pakistan (Thar Desert) is of this type. Other important deserts in this belt are the Sonoran Desert of the southwestern United States and northern Mexico and the Great Australian Desert.

Important climatic differences exist between interior and coastal deserts at the same latitude. Cold ocean currents upwelling from great depths lie offshore along the west coasts. The cold ocean current absorbs heat from the overlying air. Fog is a persistent feature and extends inland for short distances. Deserts here are cool and have low annual range of temperatures. The Atacama Desert of Chile and the Namib Desert of coastal southwest Africa are notable examples of these cool, foggy deserts.

Because desert rainfall is unreliable, river channels and the beds of smaller streams are dry most of the time (Fig. 7.4). However, a sudden and intense downpour can cause local, brief flooding that transports large amounts of sediment. Major river channels, called **wadis** (or **arroyos** in the southwestern United States), often end in flat-floored basins having no outlet (Fig. 7.5). Here clay and silt are deposited and accumulate, along with layers of soluble salts. Shallow salt lakes occupy some of these basins. Where the lakes are temporary, they are known as **playas**. In various low places in the hot desert, ground water can be reached by digging or drilling a well. Where such water supplies are available, they are often used to irrigate agricultural plots, creating an oasis (Fig. 7.6).

Dry-desert vegetation characterizes the region. Widely dispersed xerophytic plants provide negligible ground cover. In dry periods, visible vegetation is limited to small hard-leaved or spiny shrubs, cacti, or hard grasses. Many species of small annuals may be present, but they appear only after the rare, but heavy rains have saturated the soil.

In a variety of ways, the biota of the deserts have adapted to drought, a situation made worse by drying winds and high temperatures.

Fig 7.5 Death Valley's Badwater Basin in California is the point of the lowest elevation in North America at 282 ft (86 m) below sea level. With no outlet, it accumulated runoff and sediment from the surrounding mountains. During the Pleistocene ice age, Death Valley was filled with a huge lake. Vintage travel brochure by Death Valley Hotel Company; cover illustration by Gerald Cassidy, c. 1930. Author's collection

Fig 7.6 An oasis in the Sahara Desert of Libya. Date palms are planted in a sea of dune sand. Photograph by G.H. Goudarzi, U.S. Geological Survey

Succulents resist drought by storing water inside their roots and stems, and protect themselves from evaporation by having a thick, waxy layer and no leaves. Their extensive surface root-system quickly absorbs water before it sinks into porous soil. Some desert plants, such as the creosote bush, can survive without water, while others evade drought by growing close to a constant source of water, such as an oasis. Other plants, such as mesquite and tamarisk, send down very deep roots that are able to tap into a year-round supply of moisture. Some annual plants avoid the drought by lying dormant during the period between rains. Perennials, such as ocotillo, become dormant between the rains. Once all moisture has evaporated from the soil, the plant drops its leaves and temporarily stops growing.

In the Mojave-Sonoran Deserts (American Desert) of the United States and Mexico, plants are often so large that some places have a near-woodland appearance (Fig. 7.7). They include the tree-like saguaro cactus, the prickly pear cactus, the ocotillo, creosote bush, and smoke tree.

Fig 7.7 Sonoran Desert near Tucson, Arizona. Photograph by Matthew G. Bailey

Fig 7.9 Creosote bush on light desert soil in southern Nevada. Photograph by Robert G. Bailey

Fig 7.8 Barren mountains in the Peruvian littoral desert at Santa Valley, north of Lima. Photograph by Shippee-Johnson. Image #334565, American Museum of Natural History, Library

However, much of the desert of the southwestern United States is in fact scrub, thorn scrub, savannah, or steppe grassland. Parts of these regions have no visible plants. They are made up of shifting dune sand, almost sterile salt flats, or bare mountain slopes covered by a mantle of loose rock fragments (Fig. 7.8).

Over large areas of these deserts, the regolith has no soil. On the margins, however, where enough percolating water exists to moisten the upper layers of the regolith at more or less regular intervals, and where **xerophytes**, or xerophytic plants, are concentrated, a dry type of soil is formed; it is known as the **sierozem** (Aridisol). Humus is lacking and this soil is gray in color at the surface, becoming lighter in the subsoil (Fig. 7.9).

The dominant pedogenic process is **salinization**, which produces areas of salt crust where only salt-loving plants (halophytes) can survive. Calcification is conspicuous on well-drained uplands, where encrustations and deposits of calcium carbonate (caliche) are common.

Because there is little water in the deserts, mechanical weathering, or physical disintegration of the bedrock, is more rapid than chemical weathering. When a rock, such as granite, is composed of minerals of different colors, each mineral expands and contracts at a different rate. Such rocks quickly crumble into coarse sand, which forms an abrasive agent when picked up by the wind (Fig. 7.10). The very rough and youthful appearance of the desert topography is largely the result of this process. There are three kinds of desert landform regions: the **erg**, or sandy desert (Fig. 7.11), the **hamada** deserts composed of rocky plateaus channeled by dry water courses (see Frontispiece), and basin-and-range deserts of basins surrounded by barren mountains.

Animals have adapted to the desert environment. Some very specialized forms, of which the classic example is the camel must have fresh water but are able to store considerable quantities. Other species, such as scorpions, are nocturnal, carrying on their life activities at night when it is cooler and less water is needed for temperature control. Snakes and tortoises retreat to burrows. Still others, such as kangaroo rats of the American desert, restrict their water needs to what water they can manufacture by metabolism. These animals usually have little evaporation

Fig 7.10 Granite hollowed out by windblown sand, Atacama Desert, Chile. Photograph by K. Segerstrom, U.S. Geological Survey

Fig 7.11 An erg landscape in the Gobi Desert, Mongolia. Taken during the Roy Chapman Andrews Third Asiatic Expedition, 1925. Photograph by J.B. Shackelford. Image #315830, American Museum of Natural History, Library

loss from the surface and extremely dry waste products.

The world's hot deserts have been productive. Wherever irrigation is possible, the yields of crops have been high. The crops include long-staple cotton, fruits (dates, oranges, lemons, grapefruit, limes), vegetables, grains, and alfalfa. However, such irrigation projects suffer from two undesirable side effects—salinization and

waterlogging of the soil. If irrigation water is evaporated in the soil it leaves salts behind. When these are sodium salts, and they accumulate too rapidly, they form an impervious **alkali**. When drainage is not adequate to flush out the excess sodium salts, this alkali may render desert soil completely sterile.

Like the Arctic tundra, the processes of soil development and plant succession are highly vulnerable to change from human activities. A single passage of an army tank over desert soil can dramatically alter water infiltration, soil moisture, and heat distribution, and consequently biological productivity. For example, in the Mojave Desert, lichen crust on sandy soil destroyed in tracks left by tank traffic in World War II-vintage maneuvers has not recovered in 60 years.

7.4   330 Temperate Steppe Division

Temperate steppes are areas that have a semiarid continental climatic regime in which, despite maximum summer rainfall, evaporation usually exceeds precipitation. There is too little water to support a forest and too much to create a desert. Instead these regions are dominated by grasslands, which are called *shortgrass prairie* in the central United States (Columbia Plateau, Great Plains), *steppe* in Eurasia, *pampas* in South America, and *veldt* in Africa. With more moisture, the vegetation changes to savannas, which are grasslands with enough moisture to support sparse tree growth.

Trewartha (1968) classified the climate as *BSk*. The letter *k* signifies a cool climate with at least 1 month of average temperature below 0 °C. Winters are cold and dry; summers—warm to hot (see Fig. 7.1, climate diagram for Achtuba [Volograd], Russia). Drought periods are common in this climate. With the droughts come the dust storms that blow the fertile topsoil from vast areas of plowed land being used for dry farming (Fig. 7.12).

These regions are subject to periodic climatic shifts. Thus, the drought typical of arid zones may extend well outside the normal desert

Fig 7.12 A dust storm approaching in the steppe of eastern Colorado. Photograph by Soil Conservation Service

boundaries one year, and precipitation characteristic of wetter regions may make incursions into an arid zone in the next year (Fig. 7.13). These shifts in climate become critical in areas where agriculture is carried out at the margins of humidity. In semiarid areas like the Great Plains of the United States the settlers have been devastated, because these locations are sometimes desert, sometimes humid, and sometimes a hybrid of the two.

The vegetation is steppe, sometimes called shortgrass prairie, and semi-desert (Fig. 7.14). Typical steppe vegetation consists of numerous species of short grasses that usually grow in sparsely distributed bunches. Many species of grasses and other herbs occur. Buffalograss is the typical grass of the American steppe. Other typical plants are the sunflower and locoweed. Scattered shrubs and low trees sometimes grow in the steppe; all gradations of cover are present, from semi-desert to woodland. Because ground cover is generally sparse, much soil is exposed.

The semi-desert cover is xerophytic shrub vegetation accompanied by a poorly developed herbaceous layer. Trees are generally absent. An example of semi-desert cover is the sagebrush vegetation of the middle and southern Rocky Mountain region and the Colorado Plateau in the United States.

In this climatic regime, the dominant pedogenic process is calcification, with salinization on poorly drained sites. Soils contain an excess of precipitated calcium carbonate and are rich in bases. **Brown soils** (Mollisols) are typical (see Fig. 11.9, p. 108). The soils of the semi-desert shrub are sierozems (Aridisols), with little organic content, and (occasionally) clay horizons, and (in some places) accumulations of various salts. Humus content is small because the vegetation is so sparse.

The grasslands are generally areas of plains and tablelands. Few perennial streams originate here, so that those streams which do exist have their source in other more humid areas. Crossing the grasslands, the streams lose their ability to carry sediment as volume and velocity decline. The result is considerable braiding of streams during the dry season.

The dominant animals of grasslands are the large herds of hoofed grazers, such as camels in Asia, American bison, pronghorn antelope in the United States, and several species of deer worldwide. In New Zealand and Australia, mammals are replaced by large grazing birds, such as the emu (Fig. 7.15). Much of the animal life in grasslands is found underground. Rodents—such as the prairie dog in America and hamsters of Eurasia—retreat underground to escape predators and the summer heat.

These steppes constitute the great sheep and cattle ranges of the world. The steppes of central Asia have for centuries supported nomadic populations. Wheat and, to a lesser extent, oats, rye, and barley, are dominant crops. The North American Great Plains, the Ukraine, and parts of north China are all within this region. Large areas of steppe have disappeared, replaced by farmlands (Fig. 7.16).

In the Occidental world, farmers cultivate poorer land, supporting themselves on large areas by using machinery. On the dry margins they have managed to survive by allowing the land to lie fallow for a year, so that enough moisture might be stored up to permit crop production the following year.

Fig 7.13 Climatic variations in the Great Plains of the United States during normal times and a drought in 1934. From Thornthwaite (1941), p. 182

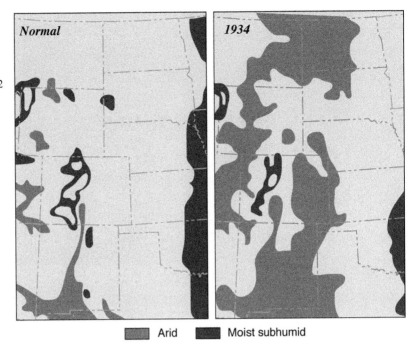

☐ Arid   ☐ Moist subhumid

Fig 7.14 Antelope herd in shortgrass prairie landscape in South Dakota. Photograph by R.S. Cole, Soil Conservation Service

7.5   340 Temperate Desert Division

Fig 7.15 The Australian emu, one of the world's largest birds, has long, powerful legs to outrun predators. It can go without water for days and withstand extreme temperatures. Drawing by Susan Strawn

Temperate deserts are located in the interior of continents, although they merge with tropical deserts equatorward. They are found in North America in the Great Basin, from Arizona, northward into Washington. In Eurasia they are found embedded in the trans-Eurasian cordillera or lie on the flanks of these mountains. Most occur in the Turkestan and Gobi regions of central Asia. In the Southern Hemisphere, only South America

Fig 7.16 Plowing with steam tractor in the steppe of North Dakota during the early 1900s. Postcard by Bloom Bros., Minneapolis, author's collection

projects far enough into mid-latitudes to a have a temperate desert climate, and only on the east side of the Andes Mountains in Patagonia, Argentina.

These desert regions have low rainfall and strong temperature contrasts between summer and winter. In the intermountain region of the western United States, between the Pacific and Rocky Mountains, the temperate desert has characteristics of a sagebrush semi-desert, with a pronounced drought season and a short humid season. Most precipitation falls in the winter, despite a peak in May (see Fig. 7.1, climate diagram of Taschkent, Uzbekistan). Aridity increases markedly in the rain shadow of the Pacific mountain ranges. Even at intermediate elevations, winters are long and cold, with temperatures below 0 °C.

These desert areas have the highest percentage of possible sunshine of any of the mid-latitude climates. Because of low humidity, over 90 % of the sun's radiation reaches the ground. These deserts experience high daily temperature fluctuations. During the summer when the days are long and the sun high in the sky, temperatures will reach 50 °C. At sunset, heat is lost rapidly because of the lack of insulating clouds. The nighttime temperature can drop over 44 °C from the daytime high, with winter temperatures in Asia's Gobi Desert plummeting to −21 °C.

Fig 7.17 Temperate semi-desert in Wyoming. Postcard by Sanborn Souvenir Co., Denver, author's collection

Under the Köppen-Trewartha system, this is the true desert, *BWk*. The letter *k* signifies that at least 1 month has an average temperature below 0 °C. These deserts differ from those at lower latitude chiefly in their far greater annual temperature range and much lower winter temperatures. Unlike the dry climates of the tropics, middle-latitude dry climates receive a portion of their precipitation as snow.

Temperate deserts support the sparse xerophytic shrub vegetation typical of semi-deserts. One example is the sagebrush vegetation of the Great Basin (Fig. 7.17) and northern Colorado Plateau region of the United States. Recently, semi-desert shrub vegetation seems to have invaded wide areas of the western United States that were formerly steppe grasslands, due to

Fig 7.18 Desert vegetation and dissected alluvial fan of the Altai Mountains, Mongolia. Taken during the Roy Chapman Andrews Third Asiatic Expedition, 1925. Photograph by J.B. Shakelford. Image #322017, American Museum of Natural History, Library

digestible foods in grasslands. Consequently, although there are herbivores in the desert, the weight in animals per unit area is small.

The surface features of these deserts consist of vast depressions surrounded by mountains or basin-and-range where flat-floored **bolsons** separate irregularly placed desert ranges. Alluvial fans surround the mountain ranges (Fig. 7.18). Water from the mountains is abundant and provides support for the large oasis communities at the base of the mountains and along the courses of exotic rivers.

overgrazing and trampling by livestock. Soils of the temperate desert are sierozems (Aridisols), low in humus and high in calcium carbonate. Poorly drained areas develop saline soils, and salt deposits cover dry lake beds.

Productivity is low due to drought and low temperatures. Growth is thus limited to a short period between winter cold and summer drought. Much of the small production goes into woody tissue. This is in contrast to the production of

References

Trewartha GT (1968) An introduction to climate, 4th edn. McGraw-Hill, New York, 408 pp

Walter H, Harnickell E, Mueller-Dombois D (1975) Climate-diagram maps of the individual continents and the ecological climatic regions of the earth. Springer, Berlin, 36 pp with 9 maps

Thornthwaite CW (1941) Climate and settlement in the Great Plains. In: Climate and man, 1941 Yearbook of agriculture. Washington, DC: U.S. Government Printing Office, pp 177–196

The Humid Tropical Ecoregions

8.1   400 Humid Tropical Domain

Equatorial and tropical air masses largely control the humid tropical group of climates found at low latitudes. Every month of the year has an average temperature above 18 °C, and no winter season. In these tropical systems, the primary periodic energy flux is diurnal: the temperature variation from day to night is greater than from season to season (see Fig. 4.2, p. 29). Average annual rainfall is heavy and exceeds annual evaporation but varies in amount, season, and distribution.

Two types of climates are differentiated on the basis of the seasonal distribution of precipitation. Figure 8.1 shows the global distribution of these two types and Fig. 8.2 shows the climographs for two stations in humid tropical climates. *Tropical wet* (or rainforest) climate has ample rainfall through 10 or more months of the year. *Tropical wet-and-dry*, or savanna, climate has a dry season more than 2 months long.

The circulation of the atmosphere controls the temporal pattern of precipitation in the tropics. Near the equator the trade winds of both hemispheres converge to form a low-pressure trough with a gentle upward drift of air. As mentioned before, this convergence zone is often referred to as the intertropical convergence zone, or ITC (see Appendix, Fig. 1, p. 142), as it represents the zone in which the trade winds from the north and south of the equator converge. Where the converging air has a trajectory over the ocean, it contains large amounts of moisture,

and cloudiness and frequent precipitation are common. Daily thunderstorms and torrential downpours predominate. Poleward of the zone, the air subsides near the tropics of Cancer and Capricorn and aridity results. The ITC shifts north and south following the migration of the vertical rays of the solar energy. This migration produces the seasonal pattern of precipitation which characterizes so much of the tropical region.

The soils of the tropics include many large areas that cannot sustain continued crop cultivation and midlatitude soils. Temperature and moisture availability are high, with rapid chemical weathering. The weathered regolith is thus quite deep. Soil profiles are often 3 m or more, and evidence of chemical weathering has been found as deep as 70 m.

Under these conditions, a process of soil development called **laterization** takes place. In the process iron, aluminum, and manganese form soluble hydroxides which tend to concentrate in the topsoil (Fig. 8.3). Highly enriched layers of iron and aluminum hydroxides, known as **laterite**, stain the soils reddish. Due to rapid chemical decomposition and solution, the soils are low in mineral nutrients. The topsoil contains little of essential elements for plant growth. The soils are also low in humus, because litter decomposes quickly. Without fertilizers, these soils can sustain crops on freshly cleared areas for only 2 or 3 years, before the nutrients are exhausted and the plot abandoned. This kind of

R.G. Bailey, *Ecoregions*, DOI 10.1007/978-1-4939-0524-9_8, © Springer Science+Media, LLC 2014   81

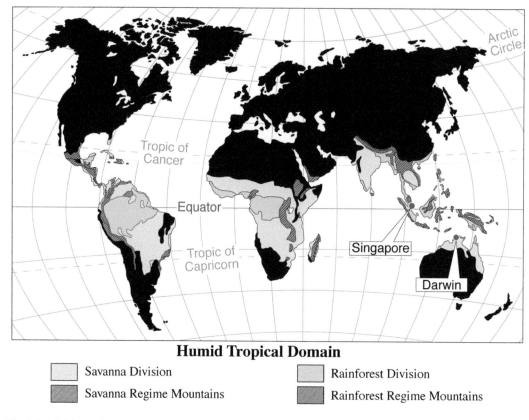

Humid Tropical Domain

	Savanna Division		Rainforest Division
	Savanna Regime Mountains		Rainforest Regime Mountains

Fig 8.1 Divisions of the continental humid tropical domain

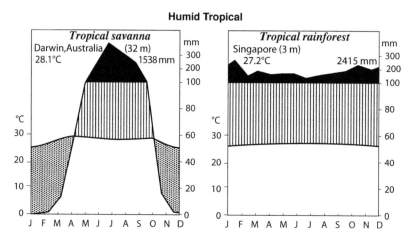

Fig 8.2 Climate diagrams of savanna and rainforest stations: (*left*), with maximum rain during the high-sun period; (*right*) with constantly wet climate. Redrawn from Walter et al. (1975)

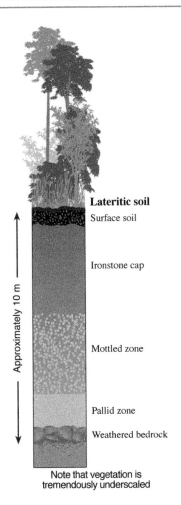

Lateritic soil

Surface soil

Ironstone cap

Mottled zone

Pallid zone

Weathered bedrock

Note that vegetation is
tremendously underscaled

Fig 8.3 Generalized profile of soil that develops in tropical climatic regimes. From King (1967), p. 175

migratory agriculture, although much reduced in areal distribution, is still one of the most important of the world's agricultural systems.

8.2   410 Savanna Division

The latitude belt between 10° and 30°N is intermediate between the equatorial and middle-latitude climates. This produces the tropical wet-dry savanna climate, which has a wet season controlled by moist, warm, maritime, tropical air masses at times of high sun, and a dry season controlled by the continental tropical masses at times of low sun (see Fig. 8.2, diagram for Darwin, Australia). Trewartha (1968) classified

Fig 8.4 Scrub woodland near Mysore in southern India. Photograph by John S. Shelton; from the University of Washington Libraries, Special Collections, John Shelton Collection, KC144-17. Reproduced with permission

the tropical wet-dry climate as *Aw*, the latter *w* signifying a dry winter.

Alternating wet and dry seasons result in the growth of three distinctive vegetation types. In some areas no grass grows at all, in others only grass grows, and in other large areas grass and trees intermingle.

The tropical scrub woodlands are made up of trees standing far enough apart so the crown of foliage fails to form a complete cover. The tree trunks are gnarled with branches all the way to the ground. The trees are deciduous, dropping all their leaves in the dry season. There are areas of low brush, called *monte* in Argentina. Where the brush is thicker and mixed with trees it forms the scrub woodland of the Chaco. In northeastern Brazil, there is an area covered by thorny, scrubby deciduous trees known as *caatinga*. Similar scrub woodlands are found in Central America and Mexico. There are also large areas in Africa, Angola, Zambia, and Zimbabwe, and also in India in the dry parts of the Deccan (Fig. 8.4). It also forms a fringe along the northern coast of Australia.

Most of the area of this type is covered with woodland savanna (Fig. 8.5). It is characterized by open expanses of tall grasses, interspersed with hardy, drought-resistant shrubs and trees. Some areas have savanna woodland, monsoon forest, thornbush, and tropical scrub. In the dry season, grasses wither into straw, and many tree species shed their leaves. Other trees and shrubs

Fig 8.5 African savanna in Kenya. The trees are acacia. Photograph by Carl Akeley. Image #211081, American Museum of Natural History, Library

Fig 8.6 The flat, featureless plain of the Orinoco Llanos in Venezuela is covered with a mixture of savanna and scrub woodland. Drawing by Susan Strawn, from photograph

have thorns and small or hard leathery leaves that resist water loss. The major areas of occurrence are in Africa, covering a large part of the low latitudes, and in the interior of Brazil.

Ribbons of dense tropical forest along most of the streams are a distinguishing characteristic of this type of region. Where the streams are small, the tall trees form a complete arch so that the streams flow through tunnels, described as galeria forests.

Plains and plateaus are the surface features most commonly found in the regions of savannas (Fig. 8.6). Some hilly uplands exist but almost no low mountain areas. Low mountains, and to a

certain, extent hilly uplands receive more rain than nearby plains or plateaus, and where sufficient rain falls, the seasonal droughts are neither too long nor too dry, so forests can survive.

Soils are mostly **latisols** (Oxisols). Heavy rainfall and high temperatures cause heavy leaching. Streamflow in these regions is subject to strong seasonal fluctuations, in striking contrast to the constant streamflow typical of rainforest climates. In the rainy season, extensive low-lying areas are submerged; in the dry season, streamflow dissipates, exposing channel bottoms of sand and gravel as stream channels and mud flats dry out.

Animal life is rich in these regions. The greatest population of herbivores and carnivores of any region on earth is found in the African zone. Africa, in particular, is so noted for its variety of species that it has been the big game-hunting center of the world.

The dry season brings a severe struggle for existence to animals of the African savanna. As streams and hollows dry up, the few muddy waterholes must supply all the drinking water (Fig. 8.7). Danger of attack by carnivores is greatly increased.

8.3   420 Rainforest Division

The wet equatorial or rainforest climate lies between the equator and latitude 10° N. Average annual temperatures are close to 27 °C; seasonal variation is imperceptible. Rainfall is heavy throughout the year, but the monthly averages differ considerably due to seasonal shifts in the intertropical convergence zone and a consequent variation in air-mass characteristics (see Fig. 8.2, climate diagram for Singapore). Trewartha (1968) defines this climate as *Ar*, with no month averaging less than 60 mm of rainfall.

The equatorial region has a forest made up of trees that cannot survive low temperatures. There are two kinds: in very wet areas with a dry season tropical rainforest grows and where a distinct dry season occurs, tropical deciduous forest grows.

The rainforest or **selva** type of vegetation is unsurpassed in number of species and luxuriant

Fig 8.7 Zebras and gnus at waterhole in Africa. Photograph by Martin Johnson. Image #314034, American Museum of Natural History, Library

Fig 8.8 Characteristic buttressing at base of cedro espinosa (*Bombacopsis quinata*) tree of the rainforest on Barro Colorado Island, Canal Zone. Image #2A4242, American Museum of Natural History, Library

tree growth (Fig. 8.8). Broadleaf trees rise 30–45 m, forming a dense leaf canopy through which little sunlight can reach the ground (Fig. 8.9). Giant lianas (woody vines) hang from trees. The forest is mostly evergreen, but individual species have various leaf-shedding cycles.

Where the moisture supply is less abundant, or where a dry season imposes a partial rhythm, a semideciduous, lighter forest grows. The dry

Fig 8.9 Rainforest canopy in Papua New Guinea. Photograph by H. Gyde Lund, U.S. Forest Service

Fig 8.10 Mangrove forest in Everglades National Park, Florida. Photograph by M.W. Williams, National Park Service

season imposes a period of rest on the winter vegetation, just as winter imposes a similar period on the forests of the middle latitudes. Many, although not all, of the trees lose their leaves. On the dry margins, the semideciduous forest merges into woodlands.

Where both the rainforest and semideciduous forest border the ocean along low, coastal plains, mangrove forest develops (Fig. 8.10). True mangrove consists of only one kind of tree of the genus *Rhizophora*. They grow on the shores of tidal swamps where the water is brackish and consists of a dense tangle of evergreen trees which grow some 5–7 m in height, with spreading bushy branches and numerous aerial roots. They serve as essential nurseries for many fish. They have been extensively cut in tropical lands to permit the entrance of ships, for making charcoal, and for aquaculture.

Tropical forests are home to small forest animals that are able to live and travel in the continuous forest canopy. These include insects, reptiles, cats, lemurs, and monkeys. Bird species are numerous and spectacularly plumaged. Underneath, in the deep shadows, are a multitude of insects, such as ants and spiders. Termites are particularly destructive of organic matter which

falls to the ground. On the forest floor there are larger mammals, including both herbivores, which graze the low leaves and fallen fruit, and carnivores, which prey on them. In Africa there are elephant, buffalo, hippopotamus, okapi, bongo, and crocodile. In the South American forest lives the anteater, which thrives on termites and ants of all kinds.

A major difference between the low-latitude rainforests and forests of higher latitude is the great diversity of species. The tropical rainforest may have as many as 1,150 different tree species in a square kilometer. The fauna of the rainforest is also very rich. A 16-km^2 area in the Canal Zone, for example, contains about 20,000 species of insects, whereas there are only a few hundred which exist in all of France.

Copious rainfall and high temperatures combine to keep chemical processes continuous on the rocks and soils. Leaching of all soluble elements of the deeply decayed rock produces red and yellow podzols (Ultisols) and latisols (Oxisols) that are often especially rich in hydroxides of iron, magnesium, and aluminum.

Streamflow is fairly constant because the large annual water surplus provides ample runoff. Dense vegetation lines river channels. Sand bars and sand banks are less conspicuous than in drier regions. Floodplains have cutoff meanders (oxbows) and many swampy sloughs where meandering river channels have shifted their courses. Although water is abundant, river systems carry relatively little dissolved material because thorough leaching of soils removes most soluble mineral matter before it reaches streams.

Not all equatorial rainforest areas have low relief. Hilly or mountainous belts have very steep slopes; frequent earthflows, slides; avalanches of soil and rock strip surfaces down to bedrock.

Man is steadily encroaching on the rainforest with lumbering, clearing for plantations, and other kinds of agriculture (Fig. 8.11). As a result, the rainforest in many places has been eliminated or decimated and replaced by scrub or open savanna, which is subject to annual fires.

Fig 8.11 Lumbering on the Island of Mindoro, Philippines. Postcard by L.J. Lambert, Manila

References

King LC (1967) The morphology of the Earth: a study and synthesis of world scenery, 2nd edn. Hafner, New York, 726 pp

Trewartha GT (1968) An introduction to climate, 4th edn. McGraw-Hill, New York, 408 pp

Walter H, Harnickell E, Mueller-Dombois D (1975) Climate-diagram maps of the individual continents and the ecological climatic regions of the earth. Springer, Berlin, 36 pp with 9 maps

The Mountain Ecoregions

9

9.1 M Mountains with Elevational Zonation

Highland systems are characterized by change more than any other regional system. On the geologic time scale they are subject to rapid changes in topography. Highland areas are associated with the margins of crustal plates, and the great elevations result from the upwarping of the crust along the plate boundaries and the upwelling of magma that forms the volcanic peaks and massive lava flows. These are the zones where volcanic activity is common and where earthquakes may be expected. The high relief, steep slopes, and generally higher precipitation accelerate erosional processes. Mass wasting is a widespread phenomenon in highlands, including avalanches and landslides (Fig. 9.1). It is these relatively high rates of geologic modification by both internal and external forces which give the highlands their rugged aspect.

Mountain climates are vertically differentiated, based on the effects of change in elevation. Air cools while ascending a mountain slope, and its capacity to hold water decreases, causing an increase in rain and snow. The thin, dry air loses heat rapidly as it ascends, and after sunset, temperatures plummet.

Elevations show typical climatic characteristics, depending on a mountain range's location in the overall pattern of global climatic zones. The climate of a given highland area is usually closely related to the climate of the surrounding lowland in seasonal character, particularly the form of the annual temperature cycle, and the times of occurrence of wet and dry seasons. For example, the Ethiopian Plateau is subject to a diurnal energy pattern, a nonseasonal energy pattern, and a seasonal moisture regime consists of a rainy summer and a dry winter (Fig. 9.2).

Since high mountains extend vertically through several climatic zones, their vegetation is usually rather sharply marked into zones. The elevational limits of various types of vegetation also correspond to the pattern of vertical temperature distribution. Roughly, the same succession of types is found in ascending mountains near the equator as it is found proceeding poleward along the continental east coasts. As discussed in Chap. 4, each mountain within a zone has a typical sequence or spectra of altitudinal belts (Table 9.1). For example, in the tropical rainforest, the tropical forest occupies the lower slopes, higher up are the mixed montane forests of broadleaf types with epiphytes (Fig. 9.3), and beyond the upper limit of trees but below the snow line, is a zone of alpine meadows. Conifers do not appear at higher elevations south of Nicaragua.

As far as vegetation and other forms of life are concerned, the vertical differentiation reaches a maximum in the low latitudes. Here we find the greatest variety of zones. Vertical differentiation remains a conspicuous feature of mountains in the middle latitudes, but disappears altogether in the high latitudes. The effect of latitude was first

R.G. Bailey, *Ecoregions*, DOI 10.1007/978-1-4939-0524-9_9, © Springer Science+Media, LLC 2014

Fig 9.1 Landslides that produce a variation in landform are widespread in highlands. From a drawing by W.M. Davis

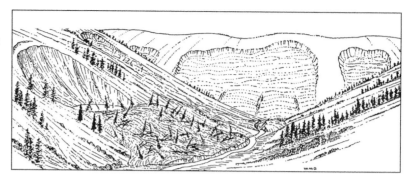

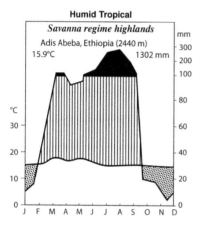

Fig 9.2 Climate diagram of Adis Abeba, 2,440 m above sea level, a tropical savanna regime highland. Data from Walter et al. (1975)

recognized by Alexander von Humboldt (1817). Much later Carl Troll developed a classification of the mountainous regions of the tropical Americas (1968), Eurasia (1972), and then expanded it worldwide (1973). The latitudinal variation in southern Rocky Mountain forests was analyzed by Peet (1978).

The position of the montane zone varies with latitude. It starts at 2,700 m in the Himalayas, 1,200 m in the Sierra Nevada of California, 900 m in the western Alps of Europe, and sea level in the Chugach Mountains of Alaska. Long, cold winters and heavy snowfall create ideal conditions for evergreen conifers, such as pines, fir, and spruce (Fig. 9.4). Food is limited in coniferous forests, and mountain animals,

such as deer and many birds migrate to higher elevations as the weather warms and food becomes abundant. In the fall, they reverse the migration. Animals not adapted to living in the cold mountain winters migrate to the lowest slopes. Year-round residents, such as bears and chipmunks, hibernate.

In mid-latitudes, the climates of these montane coniferous forests range from hot and relatively dry to cold, wet, and snowy. The nature of the forest vegetation reflects this wide climatic span. In the United States, the sparse woodland of pinyon and juniper on the desert mountains of southern California and Nevada contrast strongly with the mossy and cool spruce-fir forest of cloud-shrouded Great Smoky Mountains of North Carolina and Tennessee.

The subalpine zone is a transitional area between the lush montane forest below and the harsh alpine zone above. It is characterized by scattered, stunted, and misshapen coniferous trees, such as pines, spruces, and hemlocks. As the upper limit of forest, or tree line, is approached, the trees grow in prostrate thickets, called **krummholz**. Animals in this zone are a mixture of the animals found in the alpine zone above and the montane below. Many montane animals move to the higher elevations in summer and retreat down to more protected areas during the winter. The ibex, a goat of the European Alps, lives in the alpine area in summer and the subalpine zone in winter. Year-round residents include the yellow-bellied marmot and the alpine chipmunk.

Table 9.1 List of the types of elevational spectra

Name of division	Elevational spectra
110 Icecap	Polar desert
120 Tundra	Tundra—polar desert
130 Subarctic	Open woodland—tundra; taiga—tundra
210 Warm continental	Mixed forest—coniferous forest—tundra
220 Hot continental	Deciduous or mixed forest—coniferous forest—meadow
230 Subtropical	Mixed forest—meadow
240 Marine	Deciduous or mixed forest—coniferous forest—meadow
250 Prairie	Forest—steppe—coniferous forest—meadow
260 Mediterranean	Mediterranean woodland or shrub—mixed or coniferous forest—steppe or meadow
310 Tropical/subtropical steppe	Steppe or semi-desert—mixed or coniferous forest—alpine meadow or steppe
320 Tropical/subtropical desert	Semi-desert—shrub—open woodland—desert steppe or alpine meadow
330 Temperate steppe	Steppe—coniferous forest—tundra; steppe—mixed forest—meadow
340 Temperate desert	Semi-desert woodland—meadow
410 Tropical savanna	Open woodland—deciduous forest—coniferous forest—steppe or meadow
420 Tropical rainforest	Evergreen forest—meadow or paramos

Fig 9.3 Montane forest in the Kenya uplands. The tree limb is completely covered with epiphytes. Photograph by Carl Akeley. Image #211362, American Museum of Natural History, Library

Fig 9.4 Open montane forest of ponderosa pine in Coconino National Forest, near Happy Jack, Arizona, elevation 2,284 m. Photograph by Robert G. Bailey

If the mountain is high enough, the climate of the crest may be too severe for forests. In such places there will be a more or less marked timberline, above which is the alpine zone. The alpine zone has many extreme climatic conditions found in arctic climates, and because the atmosphere is thinner at high altitude, the light intensity is much greater. Most alpine plants grow near the ground where they are protected from the wind. Very few animals inhabit the alpine zone year-round. Those that do, such as the pika, tend to be small because of the scarcity of food. Large mammals, such as the mountain goat and the guanaco, rely on well-insulated coats for protection.

Above the tree line and the alpine zone is the climatic snow line, the boundary of the area covered by rock and permanent snow and ice (Fig. 9.5).

Other climatic elements reinforce the effects of temperature on vertical differentiation.

Fig 9.5 Snowfields above valley glaciers near Mount McKinley, Alaska. Photograph by Norman Herkenham, National Park Service

Rainfall, for instance, develops a somewhat vertical zoning. Up to 2–3 km of elevation in the middle latitudes, for example, the rainfall on mountains slopes increases. Unbroken mountain ranges are effective barriers to the passage of moisture (Fig. 9.6). The mountain ranges along the Pacific Coast of the United States, for example, intercept moisture transported from the Pacific Ocean by prevailing westerly winds, so that coastal ranges are moist and inland regions are dry. However, the effect of mountains in producing rain depends on the character of the air blowing against them. Cold air can rise only sluggishly and may produce only low stratus (sheetlike) clouds, as along the Peruvian coast (see Fig. 7.8, p. 74). The heaviest rains in the world, however, are received on mountain slopes which lie in the path of warm, buoyant, moisture-laden air, as the island of Kauai and the southern slopes of the Himalayas.

East–west mountain ranges act as temperature divides. The lowlands on the poleward side of a mountain range are made colder than they would otherwise be, and the lowlands on the equatorward side are made warmer.

Mountain climates in mid-latitudes are an important influence on river flow and floods. The higher ranges serve as snow storage areas, keeping back the precipitation until early or midsummer, releasing it slowly through melting. This maintains continuous river flow.

The upper tree line in mountainous regions is caused by the lack of adequate warmth. The lower tree line found in arid regions is related to the lack of adequate moisture. This combination restricts the growth of forests in arid regions to more or less wide bands along the slopes of mountains (Fig. 9.7).

Soils also change their character with increasing altitude, responding to the changes in climate and vegetation. Figure 9.8 shows how soil profiles change with life zones in the western United States.

Man has caused many changes in the mountain ecosystems by mining, agriculture, grazing, and fire. The elevational limits of the various forms of human settlement are similar to the horizontal limits of the lowlands. The highest settlements are associated with mining. The next highest settlements are commonly supported by the pasturage of domestic animals. Lower down, the various types of agricultural settlement appear. Both the animals and the crops supporting these settlements show fairly distinct upper limits in any one region. Because sheep can exist on much scantier pasturage than cattle, they are driven highest in the mountains. Cattle and horses usually occupy the richer pastures lower down. Of the crops, the potato reaches the highest altitude. Lower down, the various grains arrange themselves in the expectable sequence: barley, rye, wheat, maize, and rice, in descending order. The tropical crops occupy the lower slopes of low-latitude mountains.

In many areas where agriculture has been the main objective, it has been a disaster for the ecosystem. The hill lands of southern China were stripped of their once-rich soils by careless land use. The same has happened in southern Europe, in the Andes, and in the Appalachians and Ozarks of the United States. There are areas, however, where steep slopes have been tilled, using terraces to reduce erosion, for long periods of time without substantial loss. The best

Fig 9.6 Annual precipitation correlates well with the elevation of the land in an east–west section at about 38°N in the western United States. From Bailey (1941), p. 192

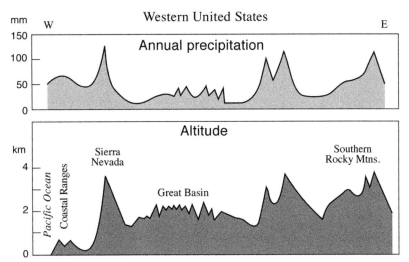

Fig 9.7 Elevational zonation in the Ruby Mountains in Nevada, with sagebrush semi-desert in the foreground, coniferous forest on the lower mountain slopes, and alpine tundra toward the top. Photograph by Robert G. Bailey

examples are found in the Himalayas, the Philippines, and the Andes.

In middle latitudes, with marked seasonal change and fewer vertical zones, mountain people tend to establish their homes at the lower altitude and to ascend each summer with their domestic animals. The movements up and down the mountains in response to seasonal rhythm are called **transhumance**. It occurs on nearly all the lower middle-latitude mountains of the world, except in Japan.

The incidence of forest fires has increased, despite forest-fire protection services restricting the once wide-ranging ground fires to relatively small areas. Exclusion of fires may cause changes in the composition and density of the vegetation, sometimes with disastrous consequences when fuels build up. Many subalpine areas in the Rocky Mountains are now occupied by successional lodgepole pine forest, which covers old burned areas in the spruce-fir zone. Our suppression of wildfires has extended the intervals between major fire events. These efforts have resulted in fires such as the infamous Yellowstone National Park fire in 1988. This fire not only had a different character from past natural fires but also burned a far larger area.

The majority of American forest and grassland ecosystems are adapted to fire of varying frequencies and magnitude. Fire is critical in maintaining ecological processes and biodiversity. Fire-excluded systems are susceptible to catastrophic fire and invasion by nonnative species. The cause of the problem in many areas includes more than a century of fire prevention and suppression along with increased human development in the **wildland-urban interface**. To correct this problem, planning for fire and land management must incorporate an improved understanding of **fire regimes**. The various kinds of fire regimes and how understanding fire regimes can abate the threat of fire exclusion and restore fire-adapted ecosystem, are discussed in Chap. 13.

Fig 9.8 Gradation of soils from a dry steppe-climate basin (*left*) to a cool, humid climate (*right*) ascending the west slopes of the Big Horn Mountains, Wyoming. (Note that the soil profile is an extreme exaggeration for purpose of illustration.) From Thorp (1931)

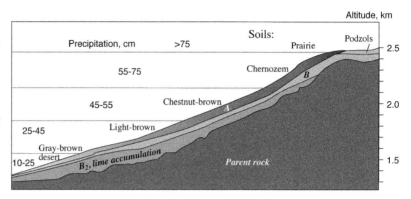

References

Bailey, R.W. 1941. Climate and settlement of the arid region. In: 1941 Yearbook of agriculture. Washington, DC: U.S. Department of Agriculture: 188–196.

Peet RK (1978) Latitudinal variation in southern Rocky Mountain forests. J Biogeography 5:275–289

Thorp J (1931) The effects of vegetation and climate upon soil profiles in northern and northeastern Wyoming. Soil Science 32:283–302

Troll C (ed) (1968) Geo-ecology of the mountainous regions of the tropical Americas. Ferd. Dummlers, Bonn, 223 pp

Troll C (ed) (1972) Geoecology of the high mountains of Eurasia. Franz Steiner, Wiesbaden, 299 pp

Troll, C. 1973. High mountain belts between the polar caps and the equator: their definition and lower limit. Arctic Alpine Res. 5(3) pt. 2: A19-A27.

Von Humboldt A (1817) De distributione geographica plantarum. Lutetiae Parisiorum, in Libraria Graeco-Latino-Germanica, Paris

Walter, H.; Harnickell, E.; Mueller-Dombois, D. 1975. Climate-diagram maps of the individual continents and the ecological climatic regions of the earth. Berlin: Springer, 36 pp with 9 maps.

Ecoregions and Climate Change

10

Ecoregions are large, region-scale ecosystems—ecoregions such as the Sonoran Desert. These regions are primarily defined by climatic conditions and on the prevailing plant formations determined by those conditions. Climate, as a source of energy and water, acts as the primary control for ecosystem distribution, including ecoregions. As climate changes, so do ecosystems, as do ancient shore lines of lakes in a desert attest (Fig. 10.1). Recognizing that climate is a principal controlling factor for ecosystems subsequently identifies the need to study the potential climatic change in terms of its ramifications to the Earth's terrestrial ecosystems. Knowing where ecological shifts will most likely occur and consequences associated with such shifts are prerequisite to productively evaluating these changes and how they affect decisions regarding resource development and management.

10.1 Long-Term Climate Change

The distribution of plant and animal communities, and indeed of entire ecoregions, has varied tremendously with past changes in climate, even in the absence of man's activities. The spatial distribution of life forms today as a function of latitude, continental position, and elevation looks very different compared to those of 5,000 or 10,000 years before the present (BP).

Climatic changes on the Earth during the past 500,000 years have been dramatic (Fig. 10.2).

Each glacial–interglacial cycle is about 100,000 years in duration, with 90,000 years of gradual climatic cooling followed by rapid warming and 10,000 years of interglacial warmth. The peak of the last glacial period, or ice age, was about 18,000 years BP and ended approximately 10,000 years BP.

During the glacial periods, the world's ice caps were greatly expanded. On the periphery of the expansive ice sheets, were correspondingly great areas of open tundra frequently underlain by permafrost. The areas of forest that form the natural vegetation of much of north and eastern North America as well as western Europe today were largely occupied by cold, rather dry tundra and steppe. A representation of the expansion and contraction of ecoclimatic zones is given in Fig. 10.3, which shows the migration of zonal belts in relation to glacial advance and retreat. Geological evidence indicates that, at certain times in the past, more water accumulated in low-latitude desert areas: huge lakes, for example, filled the now largely dry basins of the southwest United States. However, in other areas, the glacial periods were characterized, not by increased humidity, but by reduced precipitation. The most spectacular evidence for this is the great expansion of sand dunes in low latitudes: studies of air photos and satellite imagery indicate that degraded ancient dunes lie in areas that are now quite moist. Today, about 10 % of the land area between 30°N and 30°S is covered by active sand deserts. During the last great glacial advance, about 18,000 years ago,

Fig. 10.1 Shorelines of ancient Lake Bonneville, Utah. These terraces formed at various times during the ice age as the level of the lake came up and fell as the climate changed from humid to arid and back again. From G.K. Gilbert (1890)

Fig. 10.2 Changes in glacial history derived from the evidence of deep sea cores obtained from the Indian Ocean. From Imbrie and Imbrie (1979), p. 169; reproduced with permission

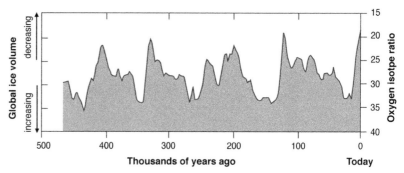

such deserts probably characterized almost 50 % of the land area at those latitudes. In this period, tropical rainforests and adjacent savannas were reduced to a narrow corridor, and were much less extensive than they are today.

Causes of such variability are evident as interpreted from the changing configuration and position of the continents and oceans (caused by **plate tectonics**), together with the uplift of mountain chains; these have strongly affected world climate. For most of the Earth's history, the continents and oceans have been so arranged that warm ocean currents from the tropics were able to flow easily into the northern and southern Polar Regions; there were no barriers to the flow of currents from low to high latitudes. This is not the situation today. At present, a continent

(Antarctica) covers a large area centered on the South Pole; the Arctic Ocean, centered on the North Pole, is almost cut off from surrounding oceans because of the arrangement of northern continents around it. Because land barriers prevent warm ocean currents from circulating the sun's energy away from tropical and temperate latitudes toward polar ones, we are now experiencing a glacial age.

Another geological change that can affect world climates is volcanic activity. A prolonged period of severe eruptions can inject large amounts of gases and particulates into the atmosphere, both of which, especially together, can decrease the quantity of solar radiation reaching the Earth's surface and so cause a phase of cooling. A meteorite that struck the Earth at the

Fig. 10.3 Suggested variability of the Earth's climatic zones over the last few hundred thousand years. From Fairbridge 1963

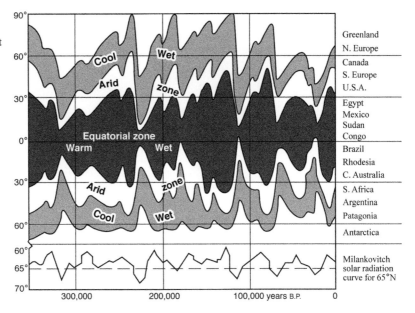

end of the Cretaceous period (65 Myr BP) is thought to have had that effect and initiated the extinction of the dinosaurs.

It is also possible that the output of solar radiation from the sun varies through time, possibly in a cyclic manner. Some good evidence derived from sunspot activity shows a pattern of 11- and 22-year cycles.

What has become more certain in recent years, however, is that the amount of radiation received at the Earth's surface through time has varied as a consequence of the Earth's ever-changing position relevant to the sun. This is called the Milankovitch cycle after its discoverer. The basic idea is that there are three ways in which the Earth's position varies (Fig. 10.4). First, the Earth's orbit around the sun is not a perfect circle but an ellipse (a). This orbital eccentricity results in approximately 3.5 % variation in the total amount of solar radiation received. Second, the tilt of the Earth's axis of rotation varies (b). And third, there is a mechanism which is based on the fact that the time of the year at which the Earth is nearest the sun varies (c). At times, the northern hemisphere is closest to the sun in winter; other times it is closest to the sun in summer. The reason for

this is that the Earth wobbles like a slowing top and swivels its axis around.

The climatic effect of these cycles is a variation in the degree of contrast between summer and winter temperatures. When the contrast between seasons is comparatively slight, summer temperatures are not high enough for the previous winter's snow and ice to melt. Snow and ice accumulate, building up huge continental ice sheets in temperate latitudes. During another phase of the cycle when there are high summer temperatures, ice melts before the onset of the succeeding winter: the result is the advent of an interglacial period, such as the one now approaching (Fig. 10.5).

10.2   Use of the Köppen Climate Classification to Detect Climate Change

The Intergovernmental Panel on Climate Change science report (IPCC 2007) represents the consensus view on **greenhouse-induced climatic changes** expressed by the overwhelming majority of atmospheric scientists throughout the world. The reported equilibrium changes for

Fig. 10.4 The three types of fluctuation in the Earth–Sun relationship involved in the Milankovitch cycle. From Christopherson, Robert W., *Geosystems: an introduction to physical geography*, 4E, © 2000, p. 517. Reprinted by permission of Pearson Education, Inc., Upper Saddle River, NJ

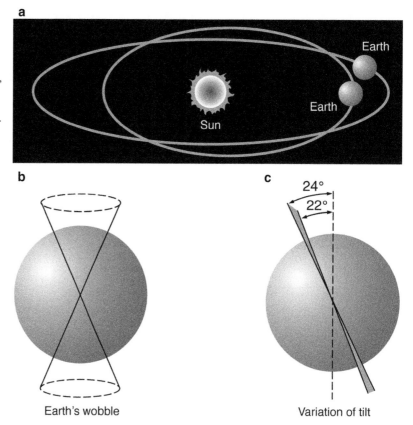

Fig. 10.5 Global climate change over the past 150,000 years and projected for the next 25,000 years. A cooling trend is projected in the future based on the Milankovitch cycles, but this may be delayed by a warming period induced by elevated concentrations of carbon dioxide and other greenhouse gases in the atmosphere. From Mitchell (1977), p. 8; in Turner et al. (2001)

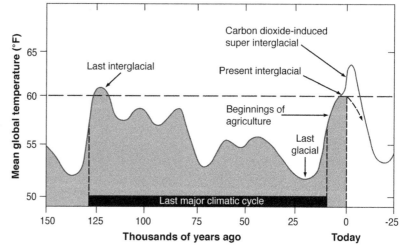

Fig. 10.6 Differences in the coverage of various Köppen climate types between periods 1961–1990 and 2036–2065 for all GCMs. From Kalvova et al. (2003)

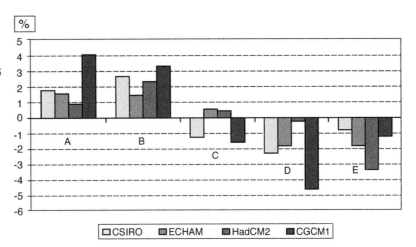

doubling of CO_2 include temperature increases between 1.5 and 4.5 °C and global precipitation increases between +3 and +15 %.

Boundaries of ecoregions coincide with certain climatic parameters. Based on macroclimatic conditions and on the prevailing plant formations determined by those conditions, I subdivided the continents into ecoregions with three levels of detail. Of these, the broadest, domains, and within them divisions, are based largely on the broad ecological zones of the German geographer, Wladimir Köppen (1931; as modified by Trewartha 1968, Chap. 4). Zone boundaries take into account the near-surface air temperature and precipitation as the major variables with respect to their annual cycles and their linkages with natural vegetation patterns. Assignment is based on quantitative definitions and, as such, can be applied to any part of the Earth where climatic data are available. It is thereby possible to develop world maps for future climate simulated, for instance, under elevated atmospheric CO_2 concentrations.

The Köppen–Trewartha classification identified six main groups of climate, and all but one—the dry group—are thermally defined (see Table 4.2). They are as follows:

Based on temperature criteria
A. Tropical: Frost limits in continental locations; in marine areas 18 °C for the coolest month

C. Subtropical: 8 months 10 °C or above
D. Temperate: 4 months 10 °C or above
E. Boreal: 1 (warmest) month 10 °C or above
F. Polar: All months below 10 °C

Based on precipitation criteria
B. Dry: Outer limits, where potential evaporation equals precipitation

Future changes in the Köppen climate types have been reported by different authors (e.g., de Castro et al. 2007; Rubel and Kottek 2010; Baker et al. 2010; Chen and Chen 2013). Gregory (1954) discusses boundary migrations in response to climate change using the Köppen–Trewartha classification. The sensitivity of the Köppen climate classification to global climatic change was tested by remapping the Köppen climate classes for an alternative climate by Kalvova et al. (2003). These investigators used the output of a series of **general circulation models** (GCM) to simulate climate for the period 1961–1990 and 2036–2065. Figure 10.6 summarizes the differences in the area distributions of the Köppen climate types. All GCM projections of warming climate (horizon 2050) show that the zones representing tropical rain climates (A) and dry climates (B) become larger, and the zones identified with boreal forest (D) and snow climates (E), together with the polar climates, are smaller. These results were similar to Lohmann et al. (1993), who did

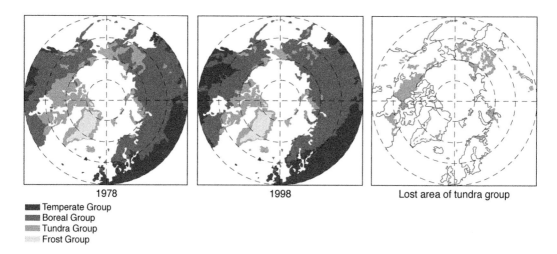

1978   1998   Lost area of tundra group

■■■ Temperate Group
■■■ Boreal Group
▨▨▨ Tundra Group
░░░ Frost Group

Fig. 10.7 Spatial distribution of Köppen tundra climate classification for selected years. From Wang and Overland (2004)

greenhouse gas warming simulations and found a retreat of regions of permafrost and the increase of areas with tropical rainy climates and dry climates.

Ecological impacts of the recent warming trend in the arctic are already noted as changes in treeline and a decrease in tundra area with the replacement of ground cover by shrubs in northern Alaska and several locations in northern Eurasia. The poleward movement of Köppen's climate zones has been documented by Wang and Overland (2004). Figure 10.7 shows the spatial distributions of Köppen's climate classifications for two selected years. The left panels of Fig. 10.7 are for 1978, the year with high tundra group coverage. By 1998, significant portions in the coverage of tundra group had been replaced by the boreal group. The coverage of tundra group being replaced by boreal group is further supported by **Normalized Differences Vegetation Index** (NDVI) data.

In response to climate change, Gao and Giorgi (2008) found a northward expansion of dry Köppen types into the Mediterranean basin and a corresponding retreat of the temperate oceanic type across the continent. They also reported a pronounced retreat of the ice cap type from the Alps. This is consistent with the worldwide retreat of glaciers expected under global warming.

Climate change may affect more than just the boundaries between ecoregions. Figure 10.8 shows the predicted elevation shift of vegetation zones in the Great Basin in Nevada (temperate desert) that would occur assuming 3 °C average climatic warming. The lower limit of woodland would shift approximately 500 m above its present elevation of 2,280 m. This would decrease the area of woodland on all mountain ranges in the region and eliminate coniferous forest from some of them. Halpin (1994) cautions that changes in ecoclimatic zonation on elevational gradients cannot be explained by simple linear assumptions applied globally. There are significant latitudinal variations in the number of elevational zones present and their elevational limits (Chap. 9). For example, the elevational limits of closed-forest timberline, tree limit, and krummholz zones vary significantly with the latitudinal position of the mountain site. There is a distinct latitudinal trend with timberlines occurring at lower elevations with distance from the equator. Conceptual models of potential impacts of climate change must take into account differences in the elevational limits of zones at different latitudes. Kupfer et al. (2005) note that increasing precipitation may tend to negate the temperature effect.

Things are further complicated by the Massenerhebung effect (German for "mountain

Fig. 10.8 The approximate elevational boundaries of the vegetation types on the isolated mountain ranges of the Great Basin: (**a**) today; (**b**) in the future after a postulated climatic warming of approximately 3 °C. From Brown, Macroecology. (c) 1995 The University of Chicago; reproduced with permission

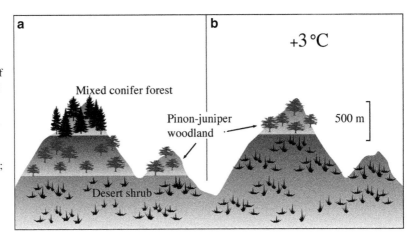

mass elevation") that described the variation in upper treeline based on the mountain size and location. In general, large mountain ranges will tend to have higher treelines than more isolated ranges because of heat retention and wind sheltering. Regions of similar elevation and latitude may have much warmer or colder climates depending on the size of the mountain ranges.

A good example of ecosystem change in both time and space is the vertical displacement of vegetation types in the Great Smoky Mountains in Tennessee and North Carolina. Using fossil pollen data, Delcourt and Delcourt (1987) show how the boreal forest in the valleys was displaced upward as the climate warmed during the past 20,000 years. In the future (2100), they project that it will be eliminated on most sites throughout the region. In its place, southern hardwood and pine forest, and temperate deciduous forest will occupy the highest elevations. Barnes (2009) provides other examples of tree response to ecosystem change.

Climate change can, in theory, cause a reduction in the spatial extent of a community. The alpine vegetation is one example of a community which will probably diminish as a direct result of climate change. Diaz and Eischeid (2007) analyzed changes in the Köppen "alpine tundra" climate classification type for the mountainous western United States by classifying 4-km pixels of topographically adjusted climate data in a geographic information system (GIS). There were 1,226 4-km pixels classified as "alpine tundra" in

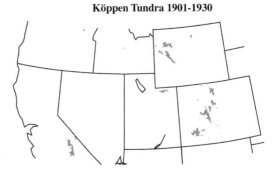

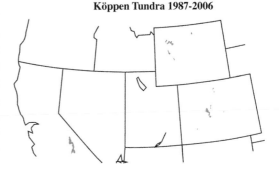

Fig. 10.9 Distribution of Köppen classification "E" (tundra climates, E-T) corresponding to the "alpine tundra" climate in the western United States. From Diaz and Eischeid (2007)

the 1901–1930 period, whereas from 1987–2006, there were only 336 thus categorized: a decline of ~73 % (Fig. 10.9). Of particular note was that the rising temperatures have caused the remaining classified alpine tundra in the last 20 years to be near the 10 °C threshold for alpine tundra classification. Continuing warming past the threshold would imply that areas where this climate type is found today in the west will no longer be present. Similar results by Shi et al. (2012) show a pronounced decrease of the cold climate types over China, including the tundra over the Tibetan Plateau.

This book uses a method to delineate regional-scale ecosystems based on the Köppen climate classification system. It should be pointed out that different methods give different predictions about the future distribution of global vegetation (and therefore ecosystem) patterns (e. g., Emanuel et al. 1985; Prentice et al. 1992; Claussen and Esch 1994; Wu et al. 2010; Zoltai 1988). It should also be noted that even if climate change and increased CO_2 does not affect the redistribution of some ecoregions, it could change their productivity and carbon storage (Parton et al. 1995).

10.3 Summary

Climate acts as the primary control for ecosystem distribution, and for the ecoregions derived from that distribution. Consequently, as the climate varies, so do the ecoregions. Climatic changes on the Earth during the past 500,000 years have been dramatic, resulting in ecoregion redistribution. Long-term climatic changes are attributable not to any single cause but to a collective set of causes: changing configuration and position of the continents (caused by plate tectonics); uplift of mountain chains; volcanic activity; output of solar radiation; and the amount of solar radiation received at the Earth's surface consequent to the Milankovitch cycles. Classifications that recognize the dependence of ecoregions on climate provide one means of constructing maps to display the impact of climate change on the geography of global ecoregions. A series of maps of the Köppen climate classification, as modified by Trewartha, was compared to Köppen map simulations based on models of climate under elevated atmospheric CO_2 concentrations. Results of studies show that the zones representing tropical rain climates and dry

climates become larger and the zones identified with boreal forest and snow climates together with the polar climates become smaller. Climate change can cause a reduction in the spatial extent of a community. Montane coniferous forest and alpine tundra are examples of communities which will probably be reduced in extent as a direct result of climate change.

References

Baker B, Diaz H, Hargrove WW et al (2010) Use of the Köppen-Trewartha climate classification to evaluate climatic refugia in statistically derived ecoregions for the People's Republic of China. Clim Change 98:113–131

Barnes BV (2009) Tree response to ecosystem change at the landscape level in eastern North America. Forstarchiv 80(3):76–89

Brown JH (1995) Macroecology. University of Chicago Press, Chicago, 269 pp

Chen D, Chen HW (2013) Using the Köppen classification to quantify climate variation and change: an example for 1901-2010. Environmental Development. http://dx.doi.org/10.1016/j.envdev.2013.03.007

Christopherson RW (2000) Geosystems: an introduction to physical geography, 4th edn. Prentice Hall, Upper Saddle River, NJ, 626 pp

Claussen M, Esch M (1994) Biomes computed from simulated climatologies. Climate Dynam 9:235–243

De Castro M, Gallardo C, Jylha K, Tuomenvirta H (2007) The use of a climate-type classification for assessing climate change effects in Europe from an ensemble of nine regional climate models. Clim Change 81:329–341

Delcourt PA, Delcourt HR (1987) Long-term forest dynamics of the temperate zone (Chapter 10). Ecological studies, vol 63. Springer, New York, pp 374–398

Diaz HF, Eischeid JK (2007) Disappearing "alpine tundra" Köppen climatic type in the western United States. Geophys Res Lett 34, L18707

Emanuel WR, Shugart HH, Stevenson MP (1985) Climatic change and the broad-scale distribution of terrestrial ecosystem complexes. Clim Change 7:29–43

Fairbridge RW (1963) Africa ice-age aridity. In: Nairn AEM (ed) Problems in paleoclimatology. Wiley, London, pp 356–363

Gao XJ, Giorgi F (2008) Increased aridity in the Mediterranean region under greenhouse gas forcing estimated from high resolution RCM simulation and the driving GCM. Global Planet Change 62:195–209

Gilbert GK (1890) Lake Bonneville. U.S. Geological Survey Monograph 1. U.S. Geological Survey, Washington, DC, 438 pp

Gregory S (1954) Climate classification and climatic change. Erdkunde 8(4):246–252

Halpin PN (1994) Latitudinal variation in the potential response of mountain ecosystems to climatic change. In: Beniston M (ed) Mountain environments in changing climates. Routledge, London, pp 180–203

Imbrie J, Imbrie KP (1979) Ice ages: solving the mystery. Enslow, Short Hills, NJ, 224 pp

IPCC (2007) Climate change 2007: the physical science basis. Contribution of working group I to the fourth assessment report of the intergovernmental panel on climate change (IPCC). Cambridge University Press, Cambridge, UK, 996 pp

Kalvova J, Halenka T, Bezpalcova K, Nemesova I (2003) Köppen climate types in observed and simulated climates. Stud Geophys Geod 47:185–202

Köppen W (1931) Grundriss der Klimakunde. Walter de Gruyter, Berlin, 388 pp

Kupfer JA, Balmat J, Smith JL (2005) Shifts in the potential distribution of Sky Island plant communities in response to climate change. In: Gottfried J, Gebow BS, Eskew LG, Edminster CB (comps.) Connecting mountain islands and desert seas: biodiversity and management of the Madrean Archipelago II. Proceedings RMRS-P-36. USDA Forest Service, Rocky Mountain Research Station, Fort Collins, CO, pp 485–490

Lohmann U, Sausen R, Bengtsson L, Cubasch U, Perlwitz J, Roeckner E (1993) The Köppen climate classification as a diagnostic tool for general circulation models. Climate Res 3:177–193

Mitchell JM (1977) Carbon dioxide and future climate. Environmental Data Service, March 1977, pp 3–9

Parton WJ, Scurlock JMO, Ojima DS, Schimel DS, Hall DO (1995) Impact of climate change on grassland production and soil carbon worldwide. Glob Chang Biol 1:13–22

Prentice IC, Cramer W, Harrison SP, Leemans R, Monserud RA, Solomon AM (1992) A global biome model based on plant physiology and dominance, soil properties and climate. J Biogeogr 19:117–134

Rubel F, Kottek M (2010) Observed and projected climate shifts 1901-2100 depicted by world maps of the Köppen-Geiger climate classification. Meteorol Z 19:135–141

Shi Y, Gao X-J, Wu J (2012) Projected changes in Köppen climate types in the 21st century over China. Atmos Ocean Sci Lett 5(6):495–498

Trewartha GT (1968) An introduction to climate, 4th edn. McGraw-Hill, New York, 408 pp

Turner MG, Gardner RH, O'Neill RV (2001) Landscape ecology in theory and practice. Springer, New York, 401 pp

Wang M, Overland JE (2004) Detecting arctic climate change using Köppen climate classification. Clim Change 67:43–62

Wu S, Zheng D, Yin Y, Lin E, Xu Y (2010) Northward-shift of temperature zones in China's eco-geographical study under future climate scenario. J Geogr Sci 20 (5):643–651

Zoltai SC (1988) Ecoclimatic provinces and man-induced climate change. Can Committee Ecol Land Classif Newslett 17:12–16

Continental Patterns and Boundaries

11

11.1 Pattern Within Zones

The zones give only a broad-brush picture. Variations within a zone break up and differentiate the major, subcontinental zones. For example, the vegetation of the savanna is highly differentiated related to variation in length of the dry season (Fig. 11.1). The geographic patterns of ecosystems within zones caused by these variations are reviewed here; see the author's *Ecosystem Geography* (Bailey 1996 et seq.) for details.

Within the same macroclimate, broad-scale landforms (geology and topography) breaks up the zonal pattern and provide a basis for further delineation of mesoscale ecosystems, known **as landscape mosaics**. The same geologic structure in different climates results in different landscapes. For example, limestone in a subarctic climate occurs in depressions and shows intense karstification, while in hot and arid climates it occurs in marked relief with a few cave tunnels and canyons inherited from colder Pleistocene time (Fig. 11.2).

A landscape mosaic may be further subdivided into microscale ecosystems called **sites**. Within a landscape, the sites are arranged in a specific pattern. For example, the Idaho Mountains, a temperate-steppe regime highland in the western United States, are made of a complex mosaic of riparian, forest, and grassland sites (Fig. 11.3).

Even in areas of uniform macroclimate, topography leads to differences in local climates and soil conditions. Topography causes variations in the amount of solar radiation received, creating **topoclimates** (Thornthwaite 1954), and affects the soil moisture (Fig. 11.4).

Variations in drainage, and in steepness of slopes, further affect the soil moisture and biota, in turn creating ecosystem sites. A sequence of moisture regimes, ranging from drier to wetter from the top to the bottom of a slope (Fig. 11.5), may be referred to as a soil catena, or a **toposequence** (Major 1951).

Figure 11.6, in a simplified way, illustrates how topography, even in areas of uniform macroclimate, leads to differences in local climates and soil conditions. The climatic climax theoretically would occur over the entire region but for topography leading to different local climates.

Other topographic, hydrologic, geologic and/or geochemical deviations may also occur. We can place ecosystem sites into three basic categories: (1) **zonal**, which are typical for the climatic conditions, such as on well-drained sagebrush terraces in a semiarid climate (Fig. 11.7); (2) **azonal** such as riparian forests and (3) **intrazonal**, which may occur on extreme types of soil that override the climatic effect, such as very dry sand dunes or black soil over certain limestone (Fig. 11.8).

In summary, the pattern of ecosystems in a region is the product of all these factors, some climatic (resulting from the average state of the atmosphere), and some **edaphic** (resulting from the character of the soil and surface). In general,

R.G. Bailey, *Ecoregions*, DOI 10.1007/978-1-4939-0524-9_11, © Springer Science+Media, LLC 2014

Fig. 11.1 Subdivision of the savannas of central Niger. From Shantz and Marbut; from *A Geography of Man*, 2nd ed., by Preston E. James, p. 304. Copyright © 1959 Ginn and Company; reprinted by permission of John Wiley & Sons, Inc.

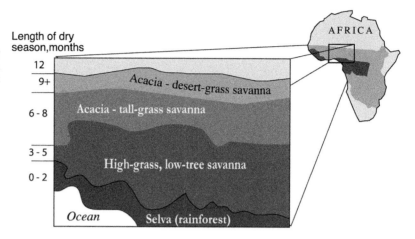

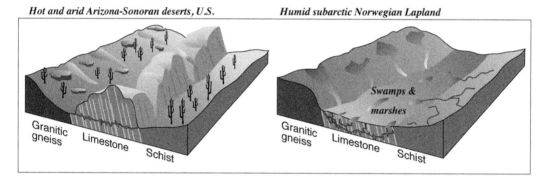

Fig. 11.2 Landscape types resulting from similar geology in two different climatic regions. From Corbel (1964)

Fig. 11.3 A mosaic site in the Idaho Mountains. Sketch by Nancy Maysmith, from photographs by John S. Shelton

the broad outline of the region is the result of climatic conditions, whereas the details observed in a particular place are the result of edaphic conditions.

Fig. 11.4 Slope and aspect affect temperature, creating topoclimates

Many smaller, natural, and man-made ecosystems are incorporated in the larger systems. These systems appear to have their own dynamics, but nevertheless are bound into the surrounding larger systems in many different ways. Three examples scattered through the land ecoregions are rivers, lakes, and towns (Fig. 11.9).

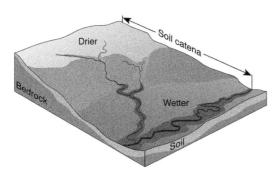

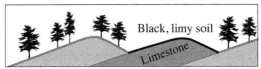

Fig. 11.8 An example of intrazonal site type where a limestone outcrop creates black, limey soil that supports grasses in the midst of a pine forest, Alabama. From Hunt (1974), p. 170

Fig. 11.5 Variation in moisture creates a toposequence or catena of soil moisture regimes

Fig. 11.6 Forest climaxes relate to topography in the temperate continental zone of southern Ontario. (Diagram is truncated, showing only three of nine possible climaxes.) Simplified from Hills (1952)

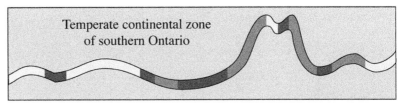

Habitat: microclimate/soil		Climax biota	
	Normal/Moist	Maple beech	Climatic climax
	Normal/Wet	Oak-ash	Edaphic Climaxes
	Normal/Dry	Oak-hickory	
	Normal/		

Fig. 11.7 Zonal sites on sagebrush terraces and azonal riparian forests in Jackson Hole, Wyoming. Photograph by National Park Service

The foregoing analysis shows how a hierarchy of spatially nested ecosystem units can be constructed by successively subdividing large ecosystems based on controlling or causal factors operating at different scales. It is about patterns created by changes in the environmental controls rather than by disturbance.

11.2   Disturbance and Succession

Disturbance and subsequent vegetation development contribute significantly to a landscape pattern at various spatial and temporal scales. An ecosystem's vegetation changes with time, and that compositional change occurs in a sequence from pioneer vegetation through successive intermediate steps to a relatively stable state called "late successional vegetation." The late successional types are used to characterize ecosystems because they tend to be far more site-specific than pioneer types, which might occur over a wider range of conditions. Furthermore,

Fig. 11.9 Man-made
ecosystem; houses sprout
from brown soil in the
semiarid steppe near Fort
Collins, Colorado.
Photograph by Robert
G. Bailey

Fig. 11.10 The northern
and western edges of the
boreal forest (tayga) in
Alaska correspond closely
to a line beyond which all
months are below 10 °C.
Climate data from Walter
and Lieth (1960–1967) and
Walter et al. (1975);
vegetation from Viereck
et al. (1992)

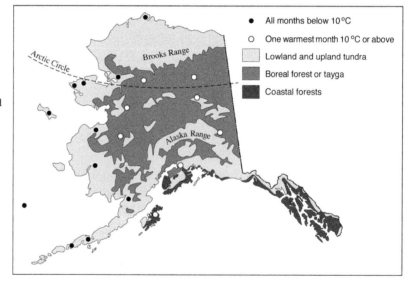

they are used as baselines for contending with the
temporal variability associated with disturbance
regimes and attending successional states of
vegetation.

11.3   Boundaries Between Zones

This scheme of defining ecosystems gives a general picture: however, the boundaries may be
imprecise. For example, the 10 °C isotherm
coincides with the northernmost limits of tree
growth; hence it separates boreal forest from
treeless tundra (Fig. 11.10).

The observer, of course, would not see an
abrupt line but a transition zone—trees on favorable sites, muskeg and bog on wetter sites, with
tundra on exposed ridges (Fig. 11.11).

For more information about ecoregion
boundaries, including 20 principles to be
observed in their delineation, see my chapter on
the subject (Bailey 2004).

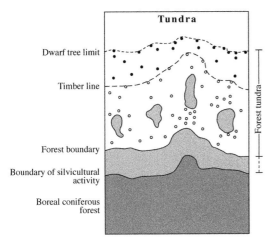

Fig. 11.11 The boundary between boreal coniferous forest and tundra is usually a transition zone rather than a sharp line. From Hustich (1953); reproduced with permission

As discussed in Chap. 10, climate exerts a very strong effect on ecoregion patterns, and climate change may cause shifts in those patterns. As the climate changes so will changes in the distribution of ecoregions as well as the boundaries between ecoregions.

References

Bailey RG (1996) Ecosystem geography. Springer, New York. 204 pp. 2 pl. in pocket

Bailey RG (2004) Identifying ecoregion boundaries. Environ Manage 34(Suppl 1):S14–S26

Corbel J (1964) L'erosion terrestre etude quantitative (methodes-technques-resultats). Ann Geogr 73:385–412

Hills A (1952) The classification and evaluation of site for forestry. Research Report 24. Ontario Department of Lands and Forest, Toronto, 41 pp

Hunt CB (1974) Natural regions of the United States and Canada. W.H. Freeman, San Francisco, 725 pp

Hustich I (1953) The boreal limits of conifers. Arct J 6:149–162

James PE (1959) A geography of man, 2nd edn. Ginn, Boston, 656 pp

Major J (1951) A functional, factorial approach to plant ecology. Ecology 32:392–412

Thornthwaite CW (1954) Topoclimatology. In: Proceedings of the Toronto meteorological conference, Royal Meterological Society, Toronto, 9–15 Sept 1953, pp 227–232

Viereck LA, Dyrness CT, Batten AR, Wenzlick KL (1992) The Alaska vegetation classification. General Technical Report PNW-GTR-286, USDA Forest Service, Pacific Northwest Research Station, Portland, OR, 278 pp

Walter H, Lieth H (1960–1967) Klimadiagramm weltatlas. G. Fischer, Jena. Maps, diagrams, profiles. Irregular pagination

Walter H, Harnickell E, Mueller-Dombois D (1975) Climate-diagram maps of the individual continents and the ecological climatic regions of the earth. Springer, Berlin. 36 pp. with 9 maps

This method of understanding processes and resultant patterns provides important design inspiration for sampling networks and managed landscapes that are **sustainable**, as well as their relevance to ecosystem management and research. These applications are reviewed here; see the author's *Ecoregion-Based Design for Sustainability* (Bailey 2002) and *Research Applications of Ecosystem Patterns* (Bailey 2009a) for details, as well as Dranstad et al. (1996), Knight and Reiners (2000), Thayer (2003), van der Ryn and Cowan (1996), and Woodward (2000). A new geography text of the United States and Canada by Chris Mayda (2012) explores sustainability within the framework of ecological regions

12.1 Design for Sustainability

As outlined in the previous chapter, ecosystems recur in predictable patterns within an ecoregion thereby reflecting processes that create these patterns. Ecoregion-based analysis strives to identify and explain geographic patterns in ecosystems in terms of formative process. **Ecoregional design** is based on the assumption that the factors which shape these patterns can be used to guide planning and design of landscapes, resulting in human-built environments which are designed differently to best fit each ecoregion's unique characteristics. By working with nature's design, designers and planners can create landscapes that function sustainably like natural ecosystems.

Several steps lead toward implementing this approach.

12.1.1 Understand Ecosystem Pattern in Terms of Process

Rather than occurring randomly, local ecosystem units occur in repetitive spatial patterns within an area called an "ecoregion." These patterns reflect a formative process. For example, rocky reservoirs support pines within grasslands of the semiarid Great Plains of the central United States (Woodward 2000). The relationship between pattern and process will vary by region.

12.1.2 Use Pattern to Design Sustainable Landscapes

The natural patterns and processes of a particular region provide essential keys to the sustainability of ecosystems, and can inspire designs for landscapes that sustain themselves. To be sustainable, a designed landscape should imitate the natural ecosystem patterns of the surrounding ecoregion in which they are embedded. As we saw before, trees signify rocky reservoirs of available water on the arid Great Plains. Planting these same trees on fine-grained plain soils, with only atmospheric precipitation to sustain them would kill the trees. By working with nature's

Fig. 12.1 Distribution of
bison in North America
from 1800 to 1975. After
Ziswiler (1967), p. 2

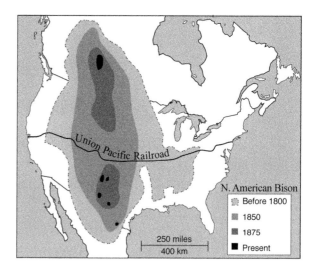

design, one can create landscapes that function
sustainably like natural ecosystems. Ecoregional
design is the act of understanding the patterns of
a region in terms of the processes that shape them
and then applying that understanding to design
and planning.

In addition:

- **Observe how a region functions and try to
  maintain functional integrity** The tropical
  rainforest, for instance, provides so much
  oxygen that it can be considered as a lung of
  the biosphere. So we should not use it only for
  massive lumbering, but instead, take advantage of its other resources, such as medicines,
  many not yet discovered. Changing the natural patterns by adding subdivisions, roads, or
  other elements changes the ecological
  functions. For example, animals change their
  routes, water flows are changed in direction
  and intensity, erosion commences, and so on.
  One of the earliest and best known examples
  of this is when the Union Pacific Railroad
  broke the large and intact habitat of the American bison into two patches separated by a
  corridor (Fig. 12.1) in 1869.
- **Maintain diversity by leaving connections
  and corridors** Fundamentally, most natural
  systems are diverse; therefore, good ecological design will maintain that diversity. Local
  ecosystems interact with each other to some

variable degree and in so doing establish some
interdependence; therefore, ecosystem diversity depends on leaving some connections and
corridors undisturbed. These principles are
being put to use in the proposed Northern
Rockies Ecosystem Protection Act (H.R.
2638). The act provides a holistic form of
ecosystem protection that explicitly connects
several of America's most beautiful
wildernesses (Fig. 12.2) and is based on the
principle that biodiversity thrives in interrelated ecosystems.

- **Honor wide-scale ecological processes** Good
  ecological design that is sustainable depends
  on honoring such natural ecological processes
  as hydrologic cycles, animal movement
  patterns, and **fire regimes**, among others.
  Identifying fire regimes will assist in fire
  planning. In the past, forest fires occurred at
  different magnitudes and frequencies in different climate-vegetation regions (Vale 1982)
  (Fig. 13.1), such as discussed in this book. In
  fire-driven ecosystems, suppressing fires or
  delaying fires indefinitely does not confer a
  sort of victory; they only assure more difficult
  fire battles in the future.
- **Match development and use to landscape
  pattern** By doing so, we allow ecological
  patterns to work for us. We can use natural
  drainage instead of storm drains, wetlands

Fig. 12.2 System of core reserves, buffer zones, and wildlife corridors proposed by the Northern Rockies Ecosystem Protection Act. Redrawn from Van der Ryn and Cowan (1996)

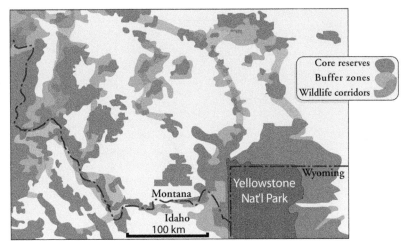

Core reserves
Buffer zones
Wildlife corridors

Wyoming

Yellowstone
Nat'l Park

Montana

Idaho
100 km

instead of sewage treatment plants, and indigenous materials rather than imported ones. Instead of channeling storm runoff into concrete drains and then to a sewage system, undeveloped drainage swales can be used to mimic nature and help provide sponges for flood protection (Barnett and Browning 1995).

- **Match development and use to the limits of the region** The solution to developing an ecological design grows from integrating design within the limits of place. For example, in the Lake Tahoe region of California-Nevada, USA (Bailey 1974), I conducted a land capability analysis using ecoregional design concepts to create land development controls that would take into account environmental limitations (e.g., soil erodibility) and ecological impacts (e.g., lake sedimentation). These controls limit land coverage (Table 12.1).

- **Design sites by considering their relationships with their neighbors** In a problem related to the Lake Tahoe site, I was to distinguish capability at both a local level and within the context of a larger area or region. My solution was to evaluate capability in two ways: on inherent features and limitations of the area; and on the geomorphic features which surround this area. This type of rating excluded small pockets of high capability lands, such as rolling uplands, when

Table 12.1 Land coverage allowances[a], Lake Tahoe Regional Planning Agency[b]

Capability district	Land coverage allowed (%)
1	1
2	1
3	5
4	20
5	25
6	30
7	35

[a]The Land Capability Map identifies the capacities of the lands in the region to withstand disturbance without risk of substantial harmful consequences occurring. These disturbances are expressed in this ordinance in terms of land coverage. Specific permitted amounts of land coverage are established for each capability district. *Source: Ordinance #12, Lake Tahoe Regional Planning Agency, p. 9*
[b]From Schneider et al. (1978)

surrounded by highly fragile, erosive, or unstable lands.

12.2   Significance to Ecosystem Management

While relevant for the design of sustainable landscapes, the concept presented above has a strong application for managing productive land uses and their environmental impacts. Understanding ecoregional patterns plays a critical role in management activities such as livestock

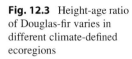

Fig. 12.3 Height-age ratio of Douglas-fir varies in different climate-defined ecoregions

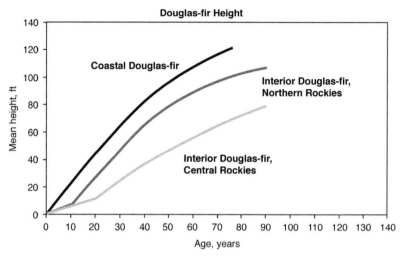

grazing, timber harvest, water diversion, and many others. An obvious application in livestock grazing is determining how much livestock for how long to maintain grazable vegetation indefinitely. Indifference to ecosystem management can lead to overgrazing that permanently diminishes an ecosystem's ability to produce grazable forage and thereby losing that ecosystem's ability to support livestock.

12.2.1 Local Systems Within Context

This perspective of seeing context can be applied in assessing the connection between action at one scale and effect at another. For example, logging on upper slopes of an ecological unit may affect downstream riparian and meadow habitats.

With the ecosystem approach, the interaction between sites can be understood because processes emerge that are not evident at the site level. An example is a snow-forest landscape that includes dark conifers that cause snow to melt faster than either a wholly snow-covered or a wholly forested basin. Landscapes function differently as a whole than would have been predicted by analysis of the individual elements (cf. Marston 2006).

The need for seeing context is also important because ecosystem characteristics have no particular regional alliance. Because of compensating factors, for example, the same forest type can occur in markedly different ecoregions: ponderosa

pine forests occur in the Northern Rockies and the southwest United States. This distribution does not imply that the climate, topography, soils, and fire regime are the same. These forests will have different productivity and response to management. For these reasons, there is a need to recognize regional differences. Cowardin et al. (1979) recommended the classification of Bailey (1976) to fill the need for regionalization for their classification of wetlands and deepwater habitats of the United States. Forest health monitoring (FHM) of the interior part of the western U.S. was conducted using an ecoregion approach to group inventory plots that have similar characteristics (Rogers et al. 2001). For the annual forest health assessments of the country, Conkling et al. (2005 et seq.) use Bailey's revised ecoregions (Cleland et al. 2007) as assessment units for analysis.

12.2.2 Spatial Transferability of Models

Predictive models differ between larger systems. The same type of forest growing in different ecoregions will occur in a different position in the landscape and have different productivity. For example, Fig. 12.3 shows that the height-age ratio of Douglas-fir varies in different climatically defined ecoregions. The ecoregion determines which ratio to apply to predict forest yield. This is important, because if a planner selects the wrong ratio, yield predictions and the forest plans upon which they are based will

Fig. 12.4 Hydrographs for three small rivers in different climate regions. Adapted from Muller and Oberlander (1978), p. 166; reproduced with permission

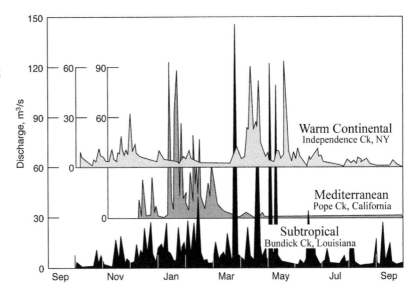

be wrong. The ecoregion map is helpful in identifying the geographic extent over which results from site-specific studies (such as **growth and yield models**) can be reliably extended. Thus the map identifies areas for the spatial transferability of models.

In Canada, studies have found that the height-diameter models of white spruce were different among different ecoregions (Huang et al. 2000). Incorrectly applying a height-diameter model fitted from one ecoregion to different ecoregions resulted in overestimation between 1 and 29 %, or underestimation between 2 and 22 %.

Another example makes an even more compelling case. Each of five regional **Forest Inventory & Analysis** (FIA) programs has developed its own set of volume models, and the models have been calibrated for regions defined by political boundaries corresponding to groups of states rather than ecological boundaries. These regional models sometimes bear little resemblance to each other. The same tree shifted a mile in various directions to move from southwest Ohio (previous Northeastern FIA) to southeast Indiana (previous North Central FIA) to northern Kentucky (Southern FIA) could have quite different model-based estimates of volume. Growth estimates are likely improved if growth models are calibrated by ecoregions rather than states or

FIA regions (Lessard et al. 2000; McNab and Keyser 2011).

Models relating **lichen** community composition in a given ecoregion to major environmental factors, such as climate and air quality, have been developed from plot data collected by FIA (Will-Wolf and Neitlich 2010; Jovan 2008).

12.2.3 Links Between Terrestrial and Aquatic Systems

Because ecoregions are based on climate and because precipitation has a climatic pattern, the streams draining any specific ecoregion have similar hydrographs (Beckinsale 1971, Fig. 12.4). This makes it possible to estimate the hydrologic productivity and streamflow characteristics of ungaged streams within the same region.

Streams depend on the terrestrial system in which they are embedded. They therefore have many characteristics in common, including biota. Delineating areas with similar climatic characteristics makes it possible to identify areas within watersheds with similar aquatic environments. A good example is the distribution of the northern hog sucker in the Ozark Uplands of Missouri, USA, which covers several watersheds (Fig. 12.5). This species of fish is widespread but not uniformly distributed throughout the Mississippi River basin. In

Fig. 12.5 Distribution of the northern hog sucker in relation to the Ozark Upland landscape and hydrologic units in Missouri. Fish data from Pflieger (1971); hydrologic unit boundaries from U.S. Geological Survey (1979)

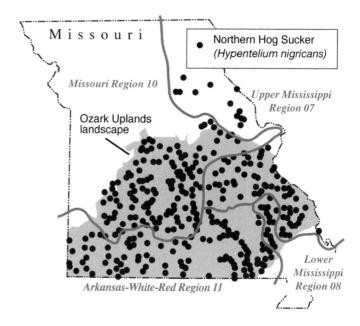

Missouri, it is found almost exclusively in the Ozark Uplands ecoregion.

12.2.4 Design of Sampling Networks

Considered collectively, the conceptual material presented to this point positions ecoregion users to design efficient sampling networks. Ecoregion maps delimit large areas of similar climate where similar ecosystems occur on similar sites. As we have seen, local ecosystems occur in predictable patterns within a particular region. Sampling representative types allow a planner, designer, or manager to extend data to analogous (unsampled) sites within the region with a high degree of reliability (Bailey 1991; Robertson and Wilson 1985), thereby reducing sampling and monitoring costs. A sampling network design should capture the local ecosystem patterns and variation in those patterns within regions exhibit variation in landform and soil characteristics (see Chap. 11). Identification of sites based on ecoregional classification could be used to **impute** their characteristics from sampled FIA sites, for example, using *k*-Nearest Neighbors or similar techniques (McRoberts et al. 2002).

Another example comes from the Rocky Mountains, a temperate steppe mountains

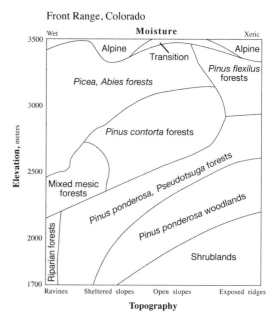

Fig. 12.6 Relationships between elevation-topography and climax plant communities, Front Range, CO. *Source*: Peet (1981) in Bailey (2009b)

ecoregion. This ecoregion, like all ecoregions, is a climatic region within which specific **plant successions** occur upon specific landform positions. The most likely successional series growing on a site within an ecoregion can be

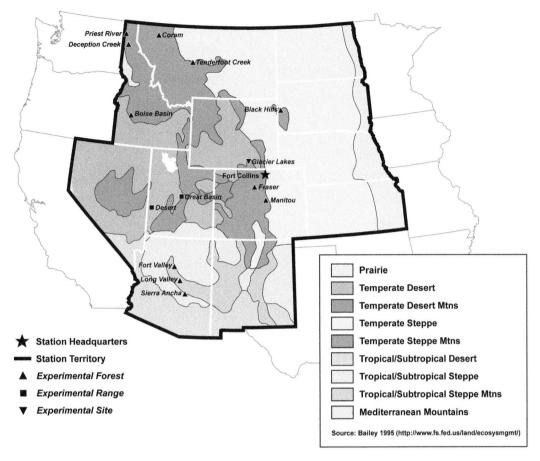

Fig. 12.7 Approximate boundaries of ecoregion divisions map (level 2 of the ecoregion hierarchy) of the U.S. Forest Service's Rocky Mountain Research Station and locations of experimental forests and ranges

predicted from landform information if the vegetation-landform relationships are known in a particular ecoregion. For example, Douglas-fir forests occur on moist, mid-elevation sites within the Front Range of Colorado. Fig. 12.6 shows the relationships between elevation-topography and climax plant communities. Understanding these relationships, vertically and horizontally, within ecoregion delineations allows the transfer of knowledge from research sites (or inventory plot) to like sites within the same ecoregion. In fact, O'Brien (1996) and Rudis (1998) found that surveys involving comprehensive sampling efforts will more accurately characterize unmonitored sites (plots) when samples are stratified according to ecologically similar areas such as ecoregions. Unfortunately, we often do not understand the

spatial relationships between the FIA plots and the landform-vegetation types within a particular ecoregion. If these were developed, we could likely produce better small-area estimates of vegetation conditions.

12.2.5 Transfer Information

Ecoregion maps show areas that are hypothesized to be analogous with respect to ecological conditions. Testing and validation of ecoregion delineations seem to bear this out (Olson et al. 1982; Inkley and Anderson 1982; Bailey 1984; McNab and Lloyd 2009). This makes it possible to transfer knowledge gained from one part of a continent to another. Figure 12.7 shows a map of ecoregions overlaid with experimental forests and ranges of the

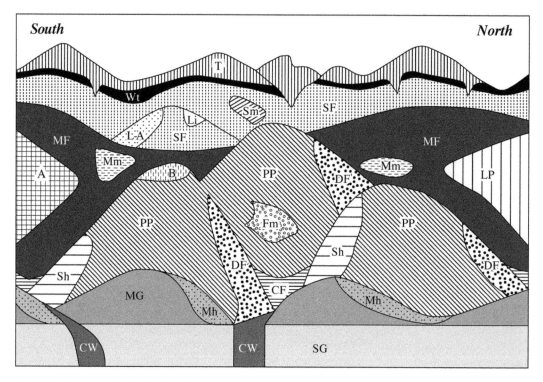

Fig. 12.8 Diagrammatic distribution of vegetation types in the mountains of the Front Range in Boulder County, CO. From Gregg (1964)

USDA Forest Service's Rocky Mountain Research Station. It shows how the ecoregion map identified forests/ranges that fall into groups with *similar ecology*. We say similar ecology because an ecoregion is a climatic region within which specific plant successions occur upon specific landform positions. The most probable vegetation growing on a site within an ecoregion can be predicted from landform information if one knows the vegetation–landform relationships in various ecoregions. (Refer to Douglas-fir example in preceding subsection.) Figure 12.8 shows the relationship between elevation-topography and climax plant communities. These relationships provide a blueprint for site analysis and management of native vegetation. Understanding the plant community gradients with respect to elevation and topography also provides a basis for separating climax from successional stands. For example, lodgepole pine forest occurring in the Douglas-fir forest zone in the Rockies may be successional following fire.

12.2.6 Determining Suitable Locations for Seed Transfer

Seed transfer zones are geographic areas within which plant materials can be moved freely with little disruption of genetic patterns or loss of local adaptation. Ecoregions have been suggested as potential seed transfer zones (Miller et al. 2011; Jones 2005) because they encompass areas with similar elevation and climate. Elevation and climate gradients appear to contribute significantly to geographic patterns of genetic variation and adaptation in many plants including trees (Post et al. 2003), shrubs, forbs, and grasses (Casler 2012).[1] One proposed refinement to the use of ecoregions as areas of plant movement has been to combine them with plant hardiness zones (Cathey 1990; revised Agricultural

---

[1] Hancock et al. (2010) found that ecoregions contribute to geographic patterns of genetic variation and adaptation of humans.

Research Service 2012) to map **plant adaptation regions** (Vogel et al. 2005). In a comparison of five region-scale ecological classification schemes, Steiner and Greene (1996) concluded that the author's ecoregion scheme was the best descriptor for regional classification of **germplasm** because of its hierarchical arrangement; the number of distinctive classes based on soils, landform, and natural vegetation; and its availability in a geographic information system format.

Ecoregions can also be used to design research. For example, Dey et al. (2009) reported that when treatment plots are located so as to account for regional differences, the results can be used to improve a manager's ability to predict oak regeneration successes and failures following given **silvicultural practices**.

12.2.7 Understanding Landscape Fragmentation

Historically, a high level of landscape heterogeneity was caused by natural disturbance and environmental gradients. Now, however, many forest landscapes appear to have been fragmented due to management activities such as timber harvesting and road construction. To understand the severity of this fragmentation, the nature and causes of the spatial patterns that would have existed in the absence of such activities should be considered. This provides insight into forest conditions that can be attained and perpetuated.

12.2.8 Choosing Planting Strategies for Landscaping and Restoration

Understanding the patterns of sites also can inspire design for urban and suburban landscapes that are in harmony with the region they are embedded within. For example, desert plants thrive on the arid south side of houses in the southwestern United States. The north side is moist and humid and can support larger, denser plants.

Furthermore, like streams, cities do not exist independently of what surrounds them. Ramage

et al. (2012) found that urban trees were consistently related to the surrounding biome (ecoregions). Classifying metropolitan areas by ecoregion forms a baseline for selecting native plants for landscaping or to restore natural conditions as well as transferring information among similar cities (Sanders and Rowntree 1983). A source of native plant information can be found in *Description of the Ecoregions of the United States* (Bailey 1995). This information is an important guide to knowing which plants will thrive in a particular ecoregion.

Gardens can be seen as extensions of the surrounding landscape and responsive to the various regions of the country. Designing urban and suburban landscapes that mimic the native vegetation by using regionally appropriate plants is the safest course to ensure landscape sustainability. By using an ecoregional pollinator guide, one can learn what native plants can be found in one's ecoregion and what pollinators they attract. These guides are published online by the Pollinator Partnership at http://www.pollinator.org/guides.htm

12.2.9 Environmental Risk Assessment

Ecological risks associated with human activity will vary depending upon the activity and where it takes place. The plants and animals of different regions respond differently to the same environmental stress. For example, Pidgeon et al. (2007) found the effect of housing development on bird species richness across the USA varied by ecoregions. Many ecoregional differences in hydrologic responses to human-modified land cover were reported by Poff et al. (2006). In the late 1970s, the ecoregion concept was used to stratify the United States into seven hydrologic regions in order to better predict the effects of silvicultural activities on **non-point source pollution** (Troendle and Leaf 1980) and later to predict the hydrologic effects of forest disturbance, including fuel reduction treatments (Troendle et al. 2010).

Hazards occur extensively in certain regions—landslides in southern California—creating a regional problem (Radbruch-Hall et al.

1982). By knowing the geographic factors that cause slides within a region, one can identify and then either avoid hazardous landslide areas or apply mitigation measures.

Likewise, certain terrestrial ecoregions have desertification risk, as their prevailing climate is arid, semi-arid, or dry subhumid, which represent 38 % of the terrestrial surface. Nunez et al. (2010) developed a method to make possible the inclusion of the desertification impact derived from human activity (agriculture, industry, mining, etc.) in land-use studies.

12.2.10 Learn from Successful Ecological Designs and Predict Establishment of Invasive Species

Ecoregion maps identify region-scale ecosystems throughout the world. For example, temperate continental ecoregions are always located in the interior of continents and on the leeward, or eastern, sides; therefore, the northeastern U.S. is ecologically similar to northern China, Korea, and Japan (Fig. 1.4, p. 3). This makes it possible to learn from successful ecological designs in similar ecoregions as well as to predict what new harmful organisms could successfully establish and spread if they were to arrive. It should be noted that not all parts of similar ecoregions are equally susceptible to the future expansion of invasive species, especially in mountain ecoregions that are broken into complex patterns of disturbance and habitats (Parks et al. 2005).

On a related note, the ecoregion concept could be useful for the safe importation of invertebrate biological control agents (Cock et al. 2006). Movement of insect species between countries in the same ecoregion is clearly less risky than moving species between disjointed similar ecoregions.

12.2.11 Maintain and Restore Biodiversity

Rather than occurring randomly, species distributions are sorted in relation to environment (Fig. 12.9). This means that similar environments tend to support similar groups of plants and animals in the absence of human disturbance (cf. Rodriguez et al. 2006).

Ecoregional analysis capitalizes on this by identifying climatic and landform factors likely to influence the distribution of species. This analysis uses these factors to define a landscape classification that groups together sites that have similar environmental character. Such a classification can then be used to indicate sites likely to have similar potential ecosystem character with similar groups of species and similar biological interactions and processes.

One of the major advantages of this approach, as opposed to directly mapping land cover, for example, is its ability to predict the potential character of sites where natural ecosystems have been profoundly modified (e.g., by land clearance or fire) or replaced by introduced plants and animals (e.g., pests and weeds).

Ecoregions have been ranked with respect to expected changes in biodiversity for the year 2100 due to climate change (Sala et al. 2000). Mediterranean climate and grassland ecosystems likely will experience the greatest proportional change in biodiversity. Northern temperate ecosystems are estimated to experience the least biodiversity change because major land-use change has already occurred.

12.2.12 Facilitate Conservation Planning

The scientific community has taken an interest in the importance of scale. Recognizing the need to move beyond traditional nature preserves to protect biodiversity, scientists have begun broadening their perspective. One of the most powerful ideas to emerge for directing conservation efforts is that of ecological regions, or ecoregions. With similar climate, geology, and landforms, ecoregions support distinctive grouping of plants and animals. Transcending unnatural political boundaries, these ecoregions provide powerful conservation planning tools.

The concept of ecoregions has been adopted by dozens of organizations in the United States and around the world as a way of thinking about

Fig. 12.9 Mammal and
plant communities on
south-facing and north-
facing slopes in lower San
Antonio Canyon, San
Gabriel Mountains,
California. From Vaughan
et al. (2000). *Mammalogy*,
4E. © Brooks/Cole, a part
of Cengage Learning, Inc.
Reproduced with
permission. www.cengage.
com/permissions

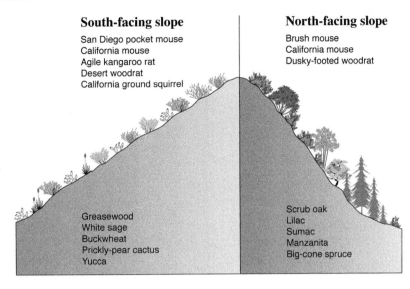

South-facing slope
San Diego pocket mouse
California mouse
Agile kangaroo rat
Desert woodrat
California ground squirrel

North-facing slope
Brush mouse
California mouse
Dusky-footed woodrat

Greasewood
White sage
Buckwheat
Prickly-pear cactus
Yucca

Scrub oak
Lilac
Sumac
Manzanita
Big-cone spruce

structuring global and continent-scale conservation efforts. For example, The Nature Conservancy has shifted the focus from conservation of single species and small sites to conservation planning on an ecoregional basis (Stein et al. 2000; Valutis and Mullen 2000). The Nature Conservancy modified Bailey's classification to identify 63 ecoregions across the United States. Organizations such as the National Wildlife Federation[2] and the U.S. Fish and Wildlife Service (cf. Corace et al. 2012) have found that ecoregions (*sensu* Bailey) define useful geographical units for conservation. Likewise the Department of Interior has initiated a national network of 22 Landscape Conservation Cooperatives (LLCs) that are based on bird conservation areas, which are loosely based on ecoregions. The World Wildlife Fund has developed an ecoregion classification system to assess the status of the world's wildlife and conserve the most biologically valuable ecoregions (Olson and Dinerstein 1998). It builds on Omernik (1987) and other analyses to provide a global-level view of ecoregions and to highlight those ecoregions worldwide that are particularly significant and should be priorities for conservation action. The U.S. Forest Service uses the Bailey ecoregion classification (Bailey 1995) to evaluate the adequacy of ecosystem representation within the National Wilderness Preservation System (Loomis and Echohawk 1999; Cordell 2012). Jepson et al. (2011) provide a critique of the various approaches to ecoregion-style conservation planning.

12.3   Significance to Research

It is important to link the ecosystem hierarchy with the research hierarchy. In so doing, research structures and ecosystem hierarchies correlate such that research information, mapping levels, and research studies work well together. Comparison of research structures and ecosystem levels can identify gaps in the research network.

At the ecoregional scale, existing research locations can be compared with ecoregion maps to identify underrepresented regions or gaps in the network (Fig. 12.10). For example, experimental forests or ranges administered by the Forest Service occur in only 26 of 52 ecoregion provinces (Lugo et al. 2006). Several ecoregions have no research facilities while others have more than one. The greatest number (14) falls within the Laurentian mixed forest ecoregion of the Lake States and Northeast. A more comprehensive analysis could include other types of

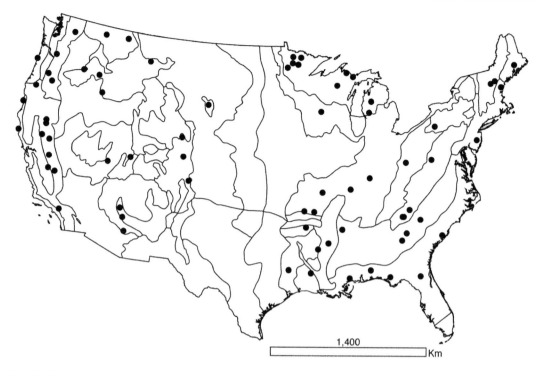

Fig. 12.10 Approximate boundaries of ecoregion provinces (level 3 of the ecoregion hierarchy) within the conterminous United States and locations of experimental forests and ranges administrated by the U.S. Forest Service. *Source*:
Alga Ramos, Forest Service, International Institute of Tropical Forestry, San Juan, Puerto Rico

similar research sites, such as Long-Term Ecological Research (LTER) sites, Long-Term
Agro-Ecosystem Research (LTAR) sites,
Research Natural Areas, National Ecological
Observatory Network (NEON) sites, and the
like. This analysis could reveal gaps in coverage
both across and within ecoregions.[3]

12.3.1 Restructuring Research Programs

The many useful applications of the study of ecosystem patterns suggest new scientific directions

for research and points the way for restructuring
research programs. To address critical ecological
issues, it is essential to move from the traditional
single-scale management and research on plots
and stands to mosaics of ecosystems (landscapes
and ecoregions) and from streams and lakes to
integrated terrestrial-aquatic systems (i.e., geographical ecosystems). FIA thematic maps (e.g.,
biomass, forest types, etc.) could assist with this.

12.3.2 Some Research Questions

These studies reveal useful applications of ecosystem patterns. There still remain many relevant
research questions associated with these patterns,
including: What are the natural ecosystem patterns
in a particular ecoregion? What are the effects of
climatic variation on ecoregional patterns and
boundaries? And, what are the relationships
between vegetation and landform in different
ecoregions? While some workers have suggested
that GIS analysis can assist in answering these

---

[3] Along similar lines, the U.S. Army has developed a
ecoregion-based map to identify environments across the
globe that are analogous to Army installations where
training and testing of soldiers and equipment take place
(Doe et al. 2000). Comparing this map with the locations
of current installations allows Army planners to assess the
ability of the Army to conduct pre-deployment activities
in similar environments, which is critical to mission
success.

questions, that approach should be used with caution because it will help identify pattern but cannot generate an understanding of the processes that create these patterns (Bailey 1988).

12.3.3 Natural Ecosystem Patterns

Historically, a high level of landscape heterogeneity was caused by natural disturbance and environmental gradients. Now, however, many forest landscapes appear to have been fragmented due to management activities such as timber harvesting and road construction. To understand the severity of this fragmentation, the nature and causes of the spatial patterns that would have existed in the absence of such activities should be considered. This analysis provides insight into forest conditions that can be attained and perpetuated (Knight and Reiners 2000).

12.3.4 Effects of Climatic Variation

Current climate exerts a very strong effect on ecosystem patterns, and climate change may cause shifts in those patterns (Neilson 1995, see Chap. 10). Anthropogenic and climatic change could yield ecoregions that are much different, or less useful, after many years. Therefore, temporal variability is an important research issue. While several researchers are doing work on the effect of climate change on tree species distribution (cf. Iverson and Prasad 2001), others are working on the impact of climatic change on the geography of ecoregions. For example, Jerry Rehfeldt of the Rocky Mountain Research Station (personal communication) has predicted the potential distribution of the American (Mojave-Sonoran) Desert ecoregion under the future climate scenario produced by the IS92a scenario of the Global Climate Model,[4] with about 21 °C warming and 50 % increase in precipitation. He has produced maps that show a greatly expanding desert under this scenario. Despite the percentage increase in precipitation, the amount of rainfall fails to keep pace with the increase in temperature, so the climate becomes more arid.

There are limits to the number of sites that can be established for monitoring changes in the global environment. Obviously, sites should be representative. Stations also should be located where they can detect change. The boundaries between climate-controlled ecoregions are suitable for this purpose. FIA has roughly 160,000 forested sample sites. This criterion could identify a subset of these sites which could be more intensively sampled to provide the needed monitoring information.

12.3.5 Relationships Between Vegetation and Landform

The relationships between vegetation and landform position change from ecoregion to ecoregion, reflecting the effect of the macroclimate. Vegetation strongly influences where animals live—some more so than others—and such factors as soil moisture and topoclimates determine which plants live where; hence site-specific vegetation character. Trees make a simple example: they change their positions in different regions (Table 12.2). Any such changes invoke related changes such as in the vigor of other tree species, ecosystem productivity, and so on. Knowledge of these differences is important for extending results of research and management experience and for designing sampling networks. These relationships have been extensively studied in some regions (cf. Odom and McNab 2000) but, unfortunately, not in others. Where sufficient studies have been done, these relationships might be modeled and mapped to improve understanding of these ecosystems.

All of the applications discussed in this chapter involve expanding our perspective to see the patterns that exist within a region. These patterns, interpreted in terms of process, can be very useful to land managers and scientists. In the next chapter, we discuss fire regimes of different ecosystems at the scale of ecoregion, and

---

[4] This is one of the emissions scenarios developed in 1992 under the sponsorship of the Intergovernmental Panel on Climate Change. IS92a has been widely adopted as a standard scenario for use in impact assessments.

Table 12.2 Relationships between vegetation and landform in various ecoregions in Ontario, Canada

Ecoregion	Topoclimate		
	Hotter	Normal	Colder
1	P		
2	P		P
3	P		P
4	A	P	P
5	A		A,P
6	C		A,P
7			C,A

From Burger (1976)

P *Picea glauca* (white spruce), A *Acer saccharum* (sugar maple), C *Carya ovata* (shagbark hickory)

go on to explore how understanding fire regimes at this scale can abate the threat of fire exclusion and restore fire-adapted ecosystems.

References

Agricultural Research Service (2012) USDA plant hardiness zone map. U.S. Department of Agriculture. Available online: http://planthardiness.ars.usda.gov

Bailey RG (1974) Land capability classification of the Lake Tahoe Basin, California-Nevada: a guide for planning. USDA Forest Service in cooperation with the Tahoe Regional Planning Agency, South Lake Tahoe, CA. With separate map at three-quarter in. equal 1 mile

Bailey RG (1976) Map: ecoregions of the United States. USDA Forest Service, Intermountain Region, Ogden, UT. Scale 1:7,500,000

Bailey RG (1984) Testing an ecosystem regionalization. J Environ Manage 19:239–248

Bailey RG (1988) Problems with using overlay mapping for planning and their implications for geographic information systems. Environ Manage 12:11–17

Bailey RG (1991) Design of ecological networks for monitoring global change. Environ Conserv 18:173–175

Bailey RG (1995) Description of the ecoregions of the United States, 2nd edn rev. and expanded (1st edn 1980). Miscellaneous Publication No. 1391 (rev.). USDA Forest Service, Washington, DC. 108 pp with separate map at 1:7,500,000

Bailey RG (2002) Ecoregion-based design for sustainability. Springer, New York, 222 pp

Bailey RG (2009a) Research applications of ecosystem patterns. In: McRoberts RE, Reams GA, Van Deuse PC, McWilliams WH (eds) Proceedings of the eighth annual forest inventory and analysis symposium. USDA Forest Service, Washington, DC, pp 83–90, 16–19 Oct 2006, Monterey, CA. General Technical Report WO-79

Bailey RG (2009b) Ecosystem geography: from ecoregions to sites, 2nd edn. Springer, New York, 251 pp

Barnett DL, Browning WD [illustrations by Uncapher JL] (1995) A primer on sustainable building. Rocky Mountain Institute, Snowmass, CO, 135 pp

Beckinsale RP (1971) River regimes. In: Chorley RJ (ed) Introduction to physical hydrology. Methuen, London, pp 176–192

Burger D (1976) The concept of ecosystem region in site classification. In: Proceedings, International Union of Forest Research Organizations (IUFRO) XVI World Congress, Division I: 20 June-2 July 1776; Oslo, Norway, pp 213–218

Casler MD (2012) Switchgrass breeding, genetics, and geonomics. In: Monti A (ed) Switchgrass: a valuable biomass crop for energy. Springer, New York, pp 29–53

Cathey HM (1990) USDA plant hardiness zone map. U.S. Department of Agriculture, Washington, DC, USDA miscellaneous publication no. 1475

Cleland DT, Freeouf JA, Keys JE et al (2007) Map: ecological subregions: sections and subsections for the conterminous United States. USDA Forest Service, Washington, DC. General Technical Report WO-76. Scale 1:3,500,000

Cock MJW, Kuhlmann U, Schaffner U, Bigler F, Babendreier D (2006) The usefulness of the ecoregion concept for safer import of invertebrate biological control agents. In: Bigler F, Babendreier D, Kuhlmann U (eds) Environmental impact of invertebrates for biological control of arthropods: methods and risk assessment. CABI Publishing, Switzerland, pp 202–221

Conkling BL, Coulston JW, Ambrose MJ (eds) (2005) Forest health monitoring, 2001 national technical report. General Technical Report SRS-81. USDA Forest Service, Southern Research Station, Asheville, NC, 204 pp

Corace RG, Shartell LM, Schulte LA et al (2012) An ecoregional context for forest management on National Wildlife Refuges of the Upper Midwest, USA. Environ Manage 49:359–371

Cordell HK (2012) The diversity of wilderness: ecosystems represented in the U.S. National Wilderness Preservation System. Int J Wilderness 18 (2):15–25, 38

Cowardin LM, Carter V, Golet FC, LaRoe ET (1979) Classification of wetlands and deepwater habitats of the United States. FWS/OBS-79/31. U.S. Fish and Wildlife Service, Washington, DC, 103 pp

Dey DC, Spetich MA, Weigel DR et al (2009) A suggested approach for design of oak (*Quercus* L.) regeneration research considering regional differences. New Forests 37:123–125

Doe WW, Shaw RB, Bailey RG, Jones DS, Macia TE (2000) U.S. Army training and testing lands: an ecoregional framework for assessment. In: Palka EJ, Galgano FA (eds) The scope of military geography: across the spectrum from peacetime to war. McGraw-Hill, New York, pp 373–392

Dranstad WE, Olson JD, Forman RTT (1996) Landscape ecology principles in landscape architecture and land-use planning. Island Press, Washington, DC, 80 pp

Gregg RE (1964) Distribution of the ant genus *Formica* in the mountains of Colorado. In: Rodeck HG (ed) Natural history of the Boulder area, vol 12. Leaflet, University of Colorado Museum, Boulder, CO, pp 59–69

Hancock AM, Witonsky DB, Ehler E et al (2010) Human adaptations to diet, subsistence, and ecoregion are due to subtle shifts in allele frequency. Proc Natl Acad Sci 107:8924–8930

Huang S, Prince C, Titus SJ (2000) Development of ecoregion-based height-diameter models for white spruce in boreal forests. For Ecol Manage 129:125–141

Inkley DB, Anderson SH (1982) Wildlife communities and land classification systems. In: Sabol K (ed) Transactions 47th North American wildlife and natural resource conference. Wildlife Management Institute, Washington, DC, pp 73–81

Iverson LR, Prasad AM (2001) Potential changes in tree species richness and forest community types following climate change. Ecosystems 4:186–199

Jepson P, Whittaker RJ, Lourie AA (2011) The shaping of the global protected area estate. In: Ladle RJ, Whittaker RJ (eds) Conservation biogeography. Blackwell Publishing, London, pp 93–125

Jones TA (2005) Genetic principles and the use of native seeds. Native Plants J 6:14–24

Jovan S (2008) Lichen bioindication of biodiversity, air quality, and climate: baseline results from monitoring in Washington, Oregon, and California. General Technical Report PNW-GTR-737. USDA Forest Service, Pacific Northwest Research Station, Portland, OR, 115 pp

Knight DH, Reiners WA (2000) Natural patterns in southern Rocky Mountain landscapes and their relevance to forest management. In: Knight DH, Smith FW, Buskirk SW, Romme WH, Baker WL (eds) Forest fragmentation in the Southern Rocky Mountains. University Press of Colorado, Boulder, CO, pp 15–30

Lessard VC, McRoberts RE, Holdaway MR (2000) Diameter growth models using FIA data from the Northeastern, Southern, and North Central Research Stations. In: McRoberts RE, Reams GA, Van Deusen PC (eds) Proceedings of the first annual forest inventory and analysis symposium. General Technical Report NC-213. USDA Forest Service, North Central Research Station, St. Paul, MN, pp 37–42

Loomis J, Echohawk JC (1999) Using GIS to identify under-represented ecosystems in the National Wilderness Preservation System in the USA. Environ Conserv 26(1):53–58

Lugo AE, Brown SL, Dodson R, Smith TS, Shugart HH (2006) Long-term research at the USDA Forest Service's Experimental Forests and Ranges. Bioscience 56(1):39–48

Marston RA (2006) President's column. Ecoregions: a geographic advantage in studying environmental change. Assoc Am Geogr Newslett 41(3):3–4

Mayda C (2012) A regional geography of the United States and Canada: toward a sustainable future. Rowman & Littlefield, Lanham, MD, 603 pp

McNab WW, Keyser CE (2011) Revisions to the 1995 map of ecological subregions that affect users of the southern variant of the Forest Vegetation Simulator. e-Research Note SRS-21. USDA Forest Service, Southern Research Station, Asheville, NC, 3 pp

McNab WH, Lloyd FT (2009) Testing ecoregions in Kentucky and Tennessee with satellite imagery and forest inventory data. In: McWilliam W et al (comps.) 2008 Forest inventory and analysis (FIA) symposium; 21–23 Oct 2008: Park City, UT. Proceedings RMRS-P-56CD. USDA Forest Service, Rocky Mountain Research Station, Fort Collins, CO. 1 CD.

McRoberts RE, Nelson MD, Wendt DG (2002) Stratified estimation of forest area using satellite imagery, inventory data, and the *k*-Nearest Neighbors technique. Remote Sens Environ 82:457–468

Miller SA, Bartow A, Gisler M et al (2011) Can an ecoregion serve as a seed transfer zone? Evidence from a common garden study with five native species. Restor Ecol 19(201):268–276

Muller RA, Oberlander TM (1978) Physical geography today: a portrait of a planet, 2nd edn. Random House, New York, 590 pp

Neilson RP (1995) A model for predicting continental-scale vegetation distribution and water balance. Ecol Appl 5:362–385

Nunez M, Civit B, Munoz P et al (2010) Assessing potential desertification environmental impact in life cycle assessment. Int J Life Cycle Assess 15:67–78

O'Brien RA (1996) Forest resources of northern Utah ecoregions. Resource Bulletin INT-RB-87. USDA Forest Service, Intermountain Research Station, Ogden, UT, 43 pp

Odom RH, McNab WH (2000) Using digital terrain modeling to predict ecological types in the Balsam Mountain of western North Carolina. Research Note SRS-8. USDA Forest Service, Southern Research Station, Asheville, NC, 11 pp

Olson DM, Dinerstein E (1998) The Global 200: a representation approach to conserving the Earth's most biologically valuable ecoregions. Conserv Biol 12:502–515

Olson RJ, Kumar KD, Burgess RL (1982) Analysis of ecoregions utilizing the geoecology data base. In: Brann TB et al (eds) In-place resource inventories: principles and practices—proceedings of a national workshop. Society of American Foresters, Bethesda, MD, pp 149–156

Omernik JM (1987) Ecoregions of the conterminous United States (map supplement). Ann Assoc Am Geogr 77:118–125

Parks CG, Radosevich SR, Endress BA, Naylor BJ, Anzinger D, Rew LJ, Maxwell BD, Dwire KA (2005) Natural and land-use history of the Northwest mountain ecoregions (USA) in relation to patterns of plant invasions. Perspect Plant Ecol Evol Syst 7:137–158

Peet RK (1981) Forest vegetation of the Colorado front range. Vegetatio 45:3–75

Pflieger WL (1971) A distributional study of Missouri fishes. University of Kansas Publication Museum of Natural History, vol 20, pp 225–570

Pidgeon AM, Radeloff VC, Flather CH et al (2007) Associations of forest bird species richness with housing and landscape patterns across the USA. Ecol Appl 17(7):1989–2010

Poff NL, Bledsoe BP, Cuhaciyan CO (2006) Hydrologic variation with land use across the contiguous United States: geomorphic and ecological consequences for stream ecosystems. Geomorphology 79:264–285

Post LS, Schlarbaum SE, van Manen F et al (2003) Development of hardwood seed zones for Tennessee using a geographic information system. South J Appl For 27(3):172–175

Radbruch-Hall DH, Colton RG, Davies WE, Lucchitta I, Skipp BA, Varmes DJ (1982) Landslide overview map of the conterminous United States. Professional Paper 1183. U.S. Geological Survey, Washington, DC. 25 pp with separate map at 1:7,500,000

Ramage BS, Roman LA, Dukes JS (2012) Relationships between urban tree communities and the biomes in which they reside. Appl Veg Sci 16(1):8–20

Robertson JK, Wilson JW (1985) Design of the national trends network for monitoring the chemistry of atmospheric precipitation. Circular 964. U.S. Geological Survey, Washington, DC, 46 pp

Rodriguez J, Hortal J, Nieto M (2006) An evaluation of the influence of environment and biogeography on community structure: the case of Holarctic mammals. J Biogeogr 33:291–303

Rogers P, Atkins D, Frank M, Parker D (2001) Forest health monitoring of the interior West. General Technical Report RMRS-GTR-75. USDA Forest Service, Rocky Mountain Research Station, Fort Collins, CO, 40 pp

Rudis VA (1998) Regional forest resource assessment in an ecological framework: the southern United States. Nat Areas J 18:319–332

Sala OE, Chapin FS III, Armesto JJ et al (2000) Global biodiversity scenarios for the year 2011. Science 287:1770–1774

Sanders RA, Rowntree RA (1983) Classification of American metropolitan areas by ecoregion and potential natural vegetation. Research Paper NE-516. USDA Forest Service, Northeastern Forest Experiment Station, Broomall, PA, 15 pp

Schneider DM, Godschalk DR, Axler N (1978) The carrying capacity concept as a planning tool. American Planning Association, Chicago, 26 pp

Stein BA, Kutner LS, Adams JS (eds) (2000) Precious heritage: the status of biodiversity in the United States. New York, Oxford University Press, 399 pp

Steiner JJ, Greene SL (1996) Proposed ecological descriptors and their utility for plant germplasm collections. Crop Sci 36:439–451

Thayer RL (2003) Life place: bioregional thought and practice. University of California Press, Berkeley, CA, 300 pp

Troendle CA, Leaf CF (1980) Hydrology. Chapter III. In: An approach to water resources evaluation of non-point silvicultural sources. U.S. Environmental Protection Agency, August 1980. EPA-600/8-8-012 Athens, GA, III.1–III.173

Troendle CA, MacDonald LH, Luce CH, Larsen IJ (2010) Fuel management and water yield (Chapter 7). In: Elliot WJ, Miller IS, Audin L (eds) Cumulative watershed effects of fuel management in the western United States. General Technical Report RMRS-GTR-231. USDA Forest Service, Rocky Mountain Research Station, Fort Collins, CO, pp 126–148

U.S. Geological Survey (1979) Accounting units of the national water data network. Washington, DC, 1:7,500,000.

Vale TR (1982) Plants and people: vegetation change in North America. Association of American Geographers Press, Washington, DC, 88 pp

Valutis L, Mullen R (2000) The Nature Conservancy's approach to prioritizing conservation action. Environ Sci Policy 3:341–346

Van der Ryn S, Cowan S (1996) Ecological design. Island Press, Washington, DC, 201 pp

Vaughan TA, Ryan JM, Czaplewski NJ (2000) Mammalogy, 4th edn. Brooks/Cole, 565 pp

Vogel KP, Schmer MR, Mitchell RB (2005) Plant adaptation regions: ecological and climatic classification of plant materials. Rangel Ecol Manage 58:351–319

Will-Wolf S, Neitlich P (2010) Development of lichen response indexes using a regional gradient modeling approach for large-scale monitoring of forests. General Technical Report PNW-GTR-807. USDA Forest Service, Pacific Northwest Research Station, Portland, OR, 65 pp

Woodward J (2000) Waterstained landscapes: seeing and shaping regionally distinctive places. Johns Hopkins University Press, Baltimore, 221 pp

Ziswiler V (1967) Extinct and vanishing animals: a biology of extinction and survival. Vol. 2 of the Heidelberg Science Library, revised English edition by F Bunnell, P Bunnell. Springer, New York, 133 pp

Use of Fire Regimes at the Ecoregion Scale

Fire-excluded ecosystems are prone to changes in composition and density and are susceptible to catastrophic fire and invasion by nonnative species. The cause of the problem in many areas includes more than a century of fire exclusion and suppression along with increased human development at the wildland-urban interface. Grazing and logging have also contributed to this problem.

To correct this problem, fire and land management must return ecosystems to a healthier, *sustainable* condition. One way to do this is to modify the current structure of ecosystems to mimic natural structures (cf., Bailey 2002).

13.1 Ecosystem Structure and Process

Ecosystem structure and process are related. For example, riparian forests evolved with flooding and fire-adapted forests evolved with fire. However, the frequency of flood or fire may vary place to place and year to year or even decade to decade; and the intensity may vary flood to flood or fire to fire. For ecosystems to sustain natural structures, they will need to experience the same kinds of processes in which they evolved (Allen et al. 2002; Savage 2003).

13.2 Range of Variation

Restoration works best if ecosystems are returned to within a "natural range of variation" (Landres et al. 1999). Ecosystems, for example, not only have variability through time because of climate change, but also across the landscape and the nation because of disturbance events, successional processes, and natural climatic variation as distinct from climate change. From forests to desert to steppe, the continent's ecosystems vary vastly. It is not possible to reconstruct how each system looked in the past. Instead, we can reset altered ecosystems back to within a range of natural variability. As Savage (2003) puts it, "If we can restore the natural processes, the natural structure should follow."

It seems reasonable that regions differing substantially in background climate should therefore have different fire regimes. In fact, fires burn with more or less regular rhythms. The simplest means to reveal a fire regime is to consider the distribution of water within an ecosystem. If they are too wet, they won't burn. The ecosystem's moisture changes with the daily and seasonal fluxes of the moisture of air masses as they move through the region. Long-term fire records around the Pacific Ocean trace nicely the pulses of the Pacific Decadal Oscillation. In this oscillation, the Pacific alternates from warm to cool phases and causes wet and dry periods on the adjacent North American continent. These wet–dry rhythms set the ecological cadence for fire regimes.

13.3   Different Ecoregions, Different Fire Regimes

Different ecoregions produce different fire regimes (Bailey 2010). Several studies have looked at variation in fire regimes at the ecoregion scale. We will examine three of them.

13.3.1 Precolonial Fire Regimes (Vale)

Precolonial fire regimes for different vegetation types in North America have been determined by analyzing fire scars. In areas lacking trees, the development of vegetation after recent fires plus early journal accounts and diaries have been used to make inferences about fire regimes. Thomas Vale (1982) synthesized this information in his book, *Plants and People*.

Vale analyzed "natural" vegetation types based on ecoregions. He used two information sources to characterize fire regimes: he extracted relevant information from 45 published fire regime studies, and he assessed species-specific fire ecology for plants indicative of the areas mapped. He found that fire regimes varied by ecoregion (Fig. 13.1). In the northern coniferous forest and woodland (boreal forest), for example, infrequent large-magnitude fires carried the flames in the canopy of the vegetation, killing most of the forest. Such fires are called "crown fires" because they burn in the upper foliage or crown of the trees.

Other environments, such as the deciduous forests of the east, probably had infrequent crown or severe surface fires. These areas are typically cool or wet and consist of vegetation that inhibits the start or spread of fire.

In mountainous regions, fire frequency is related to elevation. The lower-elevation forests in the western United States had a regime of frequent, small-magnitude, surface fires. Here,

the burning was restricted to the forest floor and most mature trees survived. Ponderosa pine forests are good examples of this kind of forest.

13.3.2 Fire Regime Types (The Nature Conservancy)

The Nature Conservancy (2004), working in cooperation with the World Wildlife Fund and the International Union for Conservation of Nature, has recently completed a global assessment of fire regime alteration on an ecoregional basis. The assessment identified three broad fire regime types (Fig. 13.2). The report reveals that, among globally important ecoregions for conservation, 84 % of the area is at risk from altered fire regimes. Almost half of priority conservation ecoregions can be classified as "fire-dependent" (shown in reddish brown).

In *fire-dependent systems*, fires are fundamental to sustaining native plants and animals. Many of the world's ecosystems—taiga forest, chaparral shrublands, savanna—have evolved with fires. What characterizes all of these ecosystems is resilience and recovery following exposure to fires. In the case of chaparral, fire does not kill most of the shrub layer; the shrubs sprout back from root crowns.

Among all the ecoregions, 36 % are *fire-sensitive*; and for this ecoregion group frequent, large and intense fires were, until recently, rare events. In these systems, plants lack the adaptations that would allow them to rapidly rebound from fire. These areas are typically cool or wet and consist of vegetation that inhibits the start or spread of fire. Examples include the tropical moist broadleaf forest and temperate rainforests.

Another 18 % of ecoregions are classified as *fire-independent ecosystems*. Here fires are largely absent because of a lack of vegetation or ignition sources, such as in Africa's Namibian Desert or in tundra ecosystems in the arctic.

According to The Nature Conservancy, fire regimes are degraded in more than 80 % of

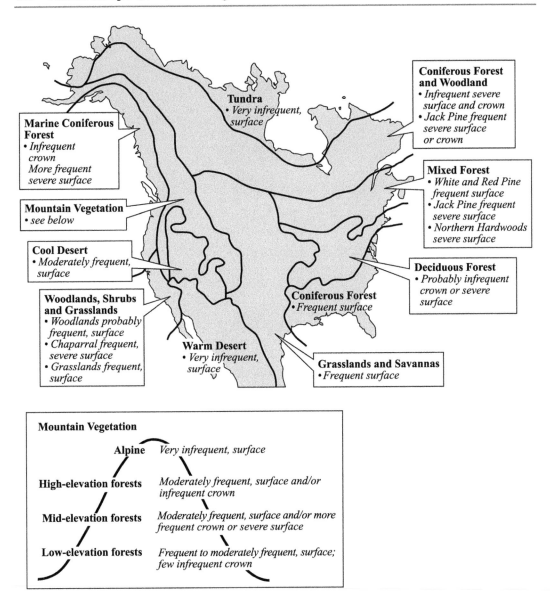

Fig. 13.1 Precolonial fire regimes of broad vegetation types (based on ecoregions) in North America. Only major divisions of the ecoregion map are shown. From Vale (1982); reprinted with permission of the Association of American Geographers

globally important ecoregions. The majority of North American forests and grasslands are adapted to fire of varying frequencies and intensities; but fire—ecologically misunderstood and therefore ecologically misinterpreted—has been suppressed so routinely for so long that some ecosystems are now likely to burn at catastrophic levels rather than at natural ecosystem-sustaining levels. Such fires favor post-fire invasion by fire-loving alien plants that once established inhibit regrowth of native vegetation, which makes recovery to the original natural ecosystem impossible. A good example is the nonnative cheatgrass that has invaded and continues to expand in sagebrush steppe in the western United States.

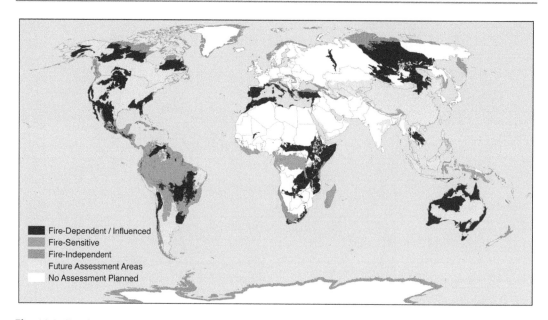

Fig. 13.2 Dominant fire regimes in priority ecoregions for biodiversity conservation. *Reddish brown* areas are fire dependent; *green* areas are fire sensitive; *gray* areas are fire independent. From The Nature Conservancy (2004)

13.3.3 Characterizing the US Wildfire Regimes (Malamud et al.)

The spread of wildfires and their severity patterns show distinct regional styles across the United States (Malamud et al. 2005). Using high-resolution Forest Service wildfire statistics, this study was based on 31 years (1970–2000) of wildfire data consisting of 88,916 fires all of which were at least 1 acre or larger and within the National Forest System. To allow spatial analysis with regard to the biophysical factors that drive wildfire regimes, the researchers classified the wildfire data into ecoregion divisions (areas of common climate, vegetation, and elevation). In each ecoregion, they asked: What is the frequency-area distribution of wildfires? The study compared area burned, number of fires, and the wildfire recurrence interval. These parameters were calculated at the ecoregion division level. The study created maps to display wildfire patterns and risk for the entire continental United States.

The authors found that the ratio of large to small wildfires decreases from east to west (Fig. 13.3a), meaning a relatively higher proportion of large fires occurs in the west compared to

the east. This may be due to greater population density and increased forest fragmentation. Alternatively, the observed gradient may be due to natural drivers, with climate, vegetation, and topography producing conditions more conducive to large wildfires.

The fire recurrence interval differs markedly between ecoregions. For example, the **fire cycle** values ranged from 13 years for the Mediterranean Mountains Ecoregion to 203 years for the Warm Continental Ecoregion (Fig. 13.3b). Note the term "fire cycle" does not mean that a fire will occur "every" 13 years, or "every" 203 years. It is a probabilistic hazard dealing with possibilities rather than absolutes: a recurrence interval of 100 years would mean that in ANY year, we have a 1 in 100 chance of a fire of a given size.

13.3.4 Other Studies

In other studies, gradients similar to those observed by Malamud et al. (2005) have been described and related to climate and vegetation. Turner and Romme (1994) describe **wildfire occurrence gradients** as a function of elevation and latitude. They attribute these gradients to

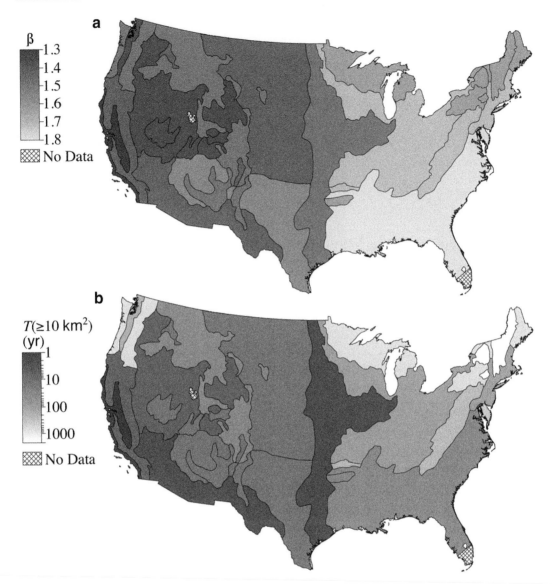

Fig. 13.3 Maps of wildfire patterns across the conterminous United States for 1970–2000 for US Forest Service wildfires classified by ecoregion division. (**a**) Ratio of large to small wildfires. The darker the color, the greater the number of large fires. (**b**) Fire recurrence interval. The legend goes from dark to light, representing "high" to "low" hazard. From Malamud et al. (2005)

broad climatic variation and note western and central regions tend to have frequent fires with forest stand structures dominated by younger trees, whereas the eastern region experiences longer inter-fire intervals and older stand structures. A statistical forecast methodology developed by Westerling et al. (2002) exploits these gradients to predict area burned in western U.S. wildfires by ecoregion a season in advance.

Littell et al. (2009) found that climate drivers of synchronous fire differ regionally. They identified four distinct geographic patterns of ecoregion provinces (ecoprovinces) across the West, and each ecoprovince had its own unique set of climate drivers that affect annual area burned by wildfire. For example, in northern mountain ecoprovinces dry, warm conditions during the seasons leading up to and including

Fig. 13.4 Percentage of increase (relative from 1950 to 2003) in median area burned in western United States ecoprovinces for a 1 °C temperature increase. From Peterson and Littell (2012)

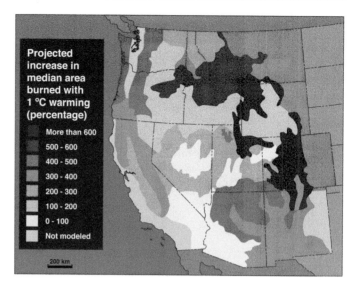

the fire season are associated with increased area burned suggesting that dry fuel condition was the key determinant of regionally synchronous fires. In contrast, in the southwestern dry ecoprovinces moist conditions the seasons prior to the fire season are more important than warmer temperatures or drought conditions in the year of the fire suggesting that fuel abundance determined large fire years.

Littell et al. (in Peterson and Littell 2012) also found that climate change will affect the area burned by ecoregion province in the western United States. They projected the statistical models of Littell et al. (2009) forward for a 1 °C temperature increase, calculated median area burned and probabilities that annual fire area would exceed the maximum annual area burned in the historical record (1950–2003). Fire area is projected to increase significantly in most ecoregion provinces (Fig. 13.4); probability of exceeding the historical maximum annual burn area varied greatly by ecoprovince. Spracklen et al. (2009) found that the forest area burned by wildfires in the western United States will increase, relative to present day, more than 50 % by 2050 as a result of climate change. The most severely affected areas will be the forests in the Pacific Northwest and Rocky Mountains ecoregion provinces, where the forest area destroyed by wildfire is predicted to increase 78 % and 178 %, respectively. Yue et al. (2013)

estimate the changes in future wildfire activity and their impact on carbonaceous aerosols over the western United States during the mid-twenty-first century will vary by ecoregion, although the estimates varied depending on which of two different fire prediction approaches was used.

Not only does fire size vary by ecoregion, it varies by land management agency. In California's Sierra Nevada region, the size of high-severity fires and percentage of high-severity fire, regardless of forest type, is less in Yosemite National Park than on Forest Service lands (Miller et al. 2012). These changes in fire regime are largely attributed to both changing climate and land management practices, including suppression of fires and past timber harvesting, during the last century. The primary Forest Service response to wildfire is contain and extinguish, while the National Park Service is more likely to let fires burn so fuels do not build up.

For a synthesis of knowledge describing how climate change will affect fire regimes at the ecoregion scale, see Sommers et al. (2011).

13.4   Use of Fire Regime at the Ecoregion Scale

The results of these studies can be used to assess burn probabilities across the nation to identify

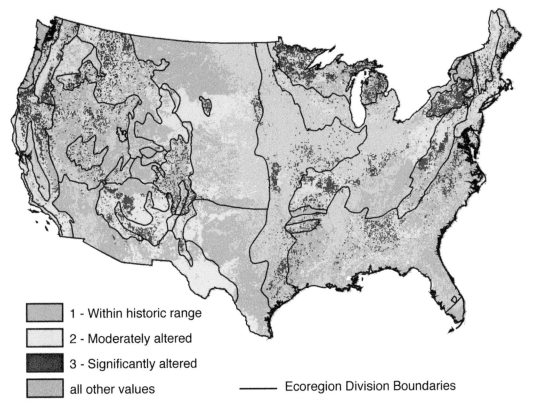

1 - Within historic range

2 - Moderately altered

3 - Significantly altered

all other values   ——— Ecoregion Division Boundaries

Fig. 13.5 Fire regime condition class (as mapped by Schmidt et al. 2002) with ecoregion division boundaries (*thick black lines*). *Green* areas (condition class 1) are largely intact and functioning; *yellow* areas (condition class 2) are moderately altered; *red* areas (condition class 3) are significantly altered; *gray* areas are nonvegetated, agricultural, or urban

areas with high risk. This helps government agencies better plan for wildfire hazards. They can also be used as a baseline from which to assess natural fire regimes, and these assessments can be used to abate the threat of fire exclusion and restore fire-adapted ecosystems. In fact, these baseline reference conditions are currently being developed as part of the LANDFIRE project (http://www.landfire.gov) by the US Forest Service (Missoula Fire Sciences Laboratory), the US Geological Survey (EROS Data Center), and The Nature Conservancy for all biophysical system across the United States. In addition, understanding fire regimes at the ecoregion scale can provide valuable insights important for designing fuel treatments by helping to determine high-hazard from low-hazard situations.

Finally, what can be done to reduce the risk of fire? Savage (2003) and Allen et al. (2002)

suggest several principles to guide the implementation of ecologically justifiable restoration projects. Two of the most important ones are:

1. Restoration of natural fire regimes (e.g., in southwestern ponderosa pine forests reduce the widespread risk of crown fires by return to low-intensity surface fire).

2. Pay attention to *both* structure and process (e.g., thinning young trees to reduce the fuel load may not work unless low-intensity surface fires are also reintroduced).

Analysis of the literature to date strongly indicates that thinning or burning treatments, or both together, do have effects consistent with restoring low-severity fire behavior in western United States pine forests (Fule et al. 2012).

Recent data from the Forest Service reflect the scale of the challenge. Schmidt et al. (2002) mapped **fire regime condition class** (FRCC), which is an ecological metric used by federal agencies, The Nature Conservancy, and others to determine the degree to which the vegetation and fire regimes of a given area have changed compared to reference conditions. As shown on their map (Fig. 13.5), fire management has significantly changed the fuel levels of many forests, and concurrently, the frequency and intensity of fire. About 30 % of all ownerships (except those related to agriculture, barren, and urban land) are in high-risk categories (shown in yellow and red). In many ecoregions this percentage is much higher. For example, in the mountains of the southwest, as much as 83 % is moderately to severely altered.

13.5   Why Ecoregions Are Needed

The same forest type can occur in different ecoregion divisions. For example, ponderosa pine forest occurs in the northern Rocky Mountains and in the southwest. This does not imply that the climate, topography, soil, and fire regime are necessarily the same. In the southwest, the historical fire regime is of frequent, low-intensity, surface fires that tend to maintain open, multi-age forests. Farther north in the Rocky Mountains, cooler conditions mean moister forests in which fires burn less readily. This distinction is important because fire management strategies and restoration protocols are often applicable only to the *local* region in which they were developed. Therefore, management strategies planned to address the fire and fuel issue such as those documented in the interagency National Fire Plan should take into consideration ecoregional variation in fire regimes. This 10-year comprehensive strategy can be viewed online at: http://www.fireplan.gov.

13.6   Use of Ecosystem Patterns Within Ecoregions

Macroclimate accounts for the largest share of systematic environmental variation at the macroscale or ecoregion level. At the mesoscale, **physiography** (geology and landform) modifies the

macroclimate and exerts the major control over ecosystem patterns and processes within climatic zones. With this in mind, Bailey et al. (1994) used physiographic factors to subdivide the ecoregion provinces of the United States into subregional areas, or *sections,* that have different landform characteristics. These differences are important because the character of the landform with different geology will vary in the climatic zone. In the same climatic zone, different geologies, such as granitic mountains or volcanic plateaus, will weather and erode differently forming different landform relief. Where this occurs, the spread of a disturbance like wildfire may differ among landforms. Swanson et al. (1990) hypothesized that in forested, steep-mountain landforms along the northwest coast of the United States where landform relief does not exceed several tree heights (e.g., Coast Ranges), disturbance agents such as fire and wind can readily move through the forest with little regard for topography. Landforms may have a greater effect on the spread of disturbance and mosaic structure where relief substantially exceeds tree height (e.g., Cascade Range). The classification and mapping of physiography as was done to delineate ecological subregions at the section level should provide an important means of discriminating broad areas with differing fire regimes within a particular ecoregion.

At finer scales, one finds considerable variation in fire regimes in response to local topography, vegetation, and microclimate (cf., Cleland et al. 2004). As we have seen, local ecosystems occur in predictable patterns within a particular ecoregion. Similar fire regimes occur on similar sites within an ecoregion. Knowledge about fire regimes on similar sites allows **ecological restoration** so as to incorporate the natural variability of fire regimes across the ecoregion.

13.7   Future Range of Variation

The range of variation concept is a useful starting point, *but* it is limited for a number of reasons. First, many systems have been fragmented because of human disturbance; because of this, fires will not carry the way they did historically.

Second, the introduction of nonnative species (e.g., cheatgrass) has made permanent changes in fire frequencies. Third, fire size and intensity of the past are clearly not acceptable in developed areas. And, fourth, system boundaries and fire regimes will change as the climate changes (cf., McKenzie et al. 1996; Rogers et al. 2011). Therefore, only where possible, we need to restore the natural range of variation. We must also determine our feasible alternatives for the "Future Range of Variation."

References

Allen CD, Savage M, Falk DA, Suckling KF, Swetnam TW, Schulke T, Stacey PB, Morgan P, Hoffman M, Klingel JT (2002) Ecological restoration of southwestern ponderosa pine ecosystems: a broad perspective. Ecol Appl 12(5):1418–1433

Bailey RG (2002) Ecoregion-based design for sustainability. Springer, New York, 222 pp

Bailey RG (2010) Fire regimes and ecoregions (Chap 2). In: Elliot WJ, Miller IS, Audin L (eds) Cumulative watershed effects of fuel management in the western United States. General technical report RMRS-GTR-231. USDA Forest Service, Rocky Mountain Research Station, Fort Collins, CO, pp 7–18

Bailey RG, Avers PE, King T, McNab WH (eds) (1994) Map. Ecoregions and subregions of the United States. U.S. Geological Survey, Washington, DC. Scale 1:7.500.000. Accompanied by supplementary table of map unit descriptions compiled and edited by W. H. McNab and R.G. Bailey

Cleland DT, Crow TR, Saunders SC, Dickmann DI, Maclean AL, Jordan JK, Watson RL, Sloan AM, Brosofske KD (2004) Characterizing historical and modern fire regimes in Michigan (USA): a landscape ecosystem approach. Landsc Ecol 19:313–325

Fule PZ, Crouse JE, Roccaforte JP, Kalies EL (2012) Do thinning and/or burning treatments in western USA ponderosa or Jeffery pine-dominated forest help restore natural fire behavior? For Ecol Manage 269:68–81

Landres PB, Morgan P, Swanson FJ (1999) Overview of the use of natural variability concepts in managing ecological systems. Ecol Appl 9(4):1379–1388

Littell JS, McKenzie D, Peterson DL, Westerling AL (2009) Climate and wildfire area burned in western U.S. ecoprovinces. 1916–2003. Ecol Appl 19(4):1003–1021

Malamud BD, Millington JDA, Perry FW (2005) Characterizing wildfire regimes in the United States. Proc Natl Acad Sci U S A 102:4694–4699

McKenzie D, Peterson DL, Alvarado E (1996) Predicting the effect of fire on large-scale vegetation patterns in North America. Research Paper PNW-RP-489. USDA Forest Service, Pacific Northwest Research Station, Portland, OR, 38 pp

Miller JD, Collins BM, Lutz JA et al (2012) Differences in wildfire among ecoregions and land management agencies in the Sierra Nevada region, California, USA. Ecosphere 3(9):80, http://dx.doi.org/10.1890/ES12-00158.1

Peterson DL, Littell JS (2012) Risk assessment for wildfire in the Western United States. In: Vose JM, Peterson DL, Patel-Weynand T (eds) Effects of climatic variability and change on forest ecosystems: a comprehensive science synthesis for the U.S. Forest sector. General Technical Report PNW-GTR-870. USDA Forest Service, Pacific Northwest Research Station, Portland, OR, pp 227–252

Rogers BM, Neilson RP, Drapek R, Lenihan JM, Wells JR, Bachelet D, Law BE (2011) Impacts of climate change on fire regimes and carbon stocks of the U.S. Pacific Northwest. J Geophys Res 116:G03037. doi:10.1029/2011JG001695

Savage M (2003) Restoring natural systems through natural processes. Quivera Coalit Newsl 6(2):1, 20–27

Schmidt KM, Menakis JP, Hardy CC, Hann WJ, Bunnell DL (2002) Development of coarse-scale spatial data for wildland fire and fuel management. General Technical Report RMRS GTR-87. USDA Forest Service, Rocky Mountain Research Station, Fort Collins, CO, 41 pp

Sommers WT, Coloff SG, Conard SG (2011) Synthesis of knowledge: fire history and climate change. Report submitted to the Joint Fire Science Program for Project 09-2-01-09, 190 pp with 6 Appendices. http://www.firescience.gov/projects/09-2-01-9/project/09-2-01-9_09_2_01_9_Deliverable_01.pdf

Spracklen DV, Mickley LJ, Logan JA, Hudman RC, Yevich R, Flannigan MD, Westerling AL (2009) Impacts of climate change from 2000 to 2050 on wildfire activity and carbonaceous aerosol concentrations in the western United States. J Geophys Res 114:D20301. doi:10.1029/2008JD010966

Swanson FJ, Franklin JF, Sedell JR (1990) Landscape patterns, disturbance, and management in the Pacific Northwest, USA. In: Zonneveld IS, Forman RTT (eds) Changing landscapes: an ecological perspective. Springer, New York, pp 191–213

The Nature Conservancy (2004) Fire, ecosystems and people. Global Fire Initiative, Tallahassee, FL, p 9

Turner MF, Romme WH (1994) Landscape dynamics in crown fire ecosystems. Landsc Ecol 9(1):59–77

Vale TR (1982) Plants and people. Association of American Geographers Press, Washington, DC, 88 pp

Westerling AL, Fershunov A, Cayon DR, Barnett TP (2002) Long-lead statistical forecasts of area burned in western U.S. wildfires by ecosystem province. Int J Wildland Fire 13:257–266

Yue X, Mickley LJ, Logan JA, Kaplan JO (2013) Ensemble projections of wildfire activity and carbonaceous aerosol concentrations over the western United States in the mid-21st century. Atmos Environ 77:767–780

Because the ideas presented in this book may be either unfamiliar or complex, or both, the seminal points deserve stressed reiteration.

1. More and more, a growing realization reveals that natural resources do not exist in isolation but interact with each other. Understanding this has led to a shifting focus to a more holistic approach of managing whole ecosystems in which there is a distinctive association of causally interconnected features, as when certain vegetation and soil types occur together with certain types of climate.

2. Ecosystems occur on many scales that are nested within each other. The boundaries are open and permeable, leading to interaction, or linkage, between systems. Ecosystems related by geography are not necessarily related by taxonomic properties. Examples include landscapes of spruce forests and glacially scoured lakes. These are not solely terrestrial or aquatic systems, but geographical ecosystem units. Because of these linkages, modification of one system affects surrounding systems. There is need to consider connections between systems, because systems are linked to form larger systems. Altering larger systems may affect smaller within. From our knowledge of the larger systems, we can much better understand the smaller systems where we must predict the outcome of land use and natural resource development.

3. The fundamental question facing all ecological mappers is: how are the boundaries of different size systems to be determined? To screen out the effects of disturbance and succession, they should be based on factors important in controlling, or causing, the ecosystem pattern at varying scales rather than what is there now so that relatively permanent boundaries can be identified to allow ecosystems to be recognized regardless of the status of the system.

4. Delineating ecosystem units involves analyzing, on a scale-related basis, the controlling factors that cause, or differentiate, ecosystem patterns at various scales, and then use significant changes in controls as boundaries. Climate, as a source of energy and water, acts as the primary control for ecosystem distribution. As the types of climate change so do the ecosystems. The most important climatic factor determining ecosystem boundaries is climatic regime, defined as the seasonality of temperature and moisture. As the climatic regime changes, so do the hydrologic and erosion cycles and the life cycles of the biota. Controls over the climatic effect change with scale. At the global level, or macro-scale, ecosystem patterns are controlled by the macroclimate related to variation in latitude and the character of the surface. Global variation in macroclimate forms the world's ecoclimatic zones, also known as

regional-scale ecosystems, or ecoregions. This book classifies and plots their distribution based on specific criteria that define what type of region each is. It provides illustrated descriptions of 31 terrestrial and 15 oceanic ecoregions in a comparative context.

5. A unique feature of this book is its coverage of both oceanic and continental ecosystems. Oceans occupy some 70 % of the Earth's surface. In a hierarchical sense, they are the environment of the continental system embedded within, controlling their behavior, through their influence on climatic patterns. Understanding continental systems requires a grasp of the enormous influence that marine systems exert on terrestrial climatic patterns and thus the character and distribution of continental ecosystems. The surface of the ocean is differentiated into regions with different hydrology, defined as the seasonal variation in temperature and salinity. Ocean hydrology controls the distribution of life in the oceans and is the basis for regional-scale ecosystem units. The book includes extensive discussion of the factors controlling ocean hydrology. Building on the work of Günter Dietrich, it focuses on the roles that latitude, wind systems, precipitation, and evaporation play in affecting the distribution of the Earth's major oceanic regions.

6. Based on macroclimate and on macro-features of the vegetation determined by those conditions, the continents are subdivided into ecoregions with three levels of detail. Their boundaries are determined using the Köppen–Trewatha climate classification system as a starting point in combination with potential natural vegetation at the level of plant formation. The arrangement of the ecological climate zones depends largely on latitude and continental position. To further complicate matters, the Earth's internal energy causes irregular patterns of high mountains on the continents. These modify the climate that would otherwise exist on a flat continent. Mountains exhibiting elevational zonation and having the climatic regime of the adjacent lowlands are distinguished according to the character of the zonation. These will differ according to the climatic zone in which they are embedded.

7. We have the knowledge base, the mapping tools, and the analytical protocols necessary to evaluate factors affecting the distribution of the Earth's major ecoregions. Our objective, therefore, should shift from mere empirical description of site-specific localities to discovering and documenting the mechanisms that are responsible for producing the world pattern of ecoregions. Understanding spatial relationships between causal mechanisms and resultant patterns is the key to understanding ecosystem dynamics and how they respond to management.

8. We can interpret the patterns of both ocean and continental ecoregions through macroclimate. Ecoregions recur in similar form in various parts of the world. Because of this predictability, we can transfer knowledge gained about one region to another; and because data can be reliably extended to analogous sites within a region, we may greatly reduce data sampling and monitoring.

9. As with the larger ecoregions, the pattern of sites within each continental region also recurs predictably; but the pattern of local ecosystems is controlled by finer scale climatic variation that result from differences in landform. Local variation in landform (geology and topography) will cause small-scale variations in the amount of solar radiation received, create topoclimates, and affect the amount of soil moisture. These variations will subsequently affect the biota, creating ecosystem sites as subdivisions of a larger ecoregion.

10. With recognition that climate is the primary controlling factor for ecosystem distribution, there exists a need to study potential climate change on ecosystem distribution, from the regional (ecoregion) to the local, site scale. Knowing where ecological shifts will most

likely occur and consequences associated with such shifts are prerequisite to the evaluation of these changes in terms of development and resource management decisions. Ecoregion maps show the Earth's surface subdivided into areas based on large patterns of ecosystems. These regions delimit large areas within which local ecosystems recur throughout the region in a predictable fashion. By observing the behavior of the different systems within a region, it is possible to predict the behavior of an unvisited one. Hence a map of this type can be used to spatially extend data obtained from limited sample sites. The results of observations at representative sites from each region would be potentially useful in detecting and monitoring climate change effects. Such a map can aid with the design of a monitoring network.

11. Different ecoregions produce different conditions for fire as a natural process. In some ecoregions fire may be infrequent, rare, or even essentially absent. In other ecoregions fire may recur in a pattern and thereby present as a natural fire regime, but fire regimes may differ even among fire-adapted ecosystems. Results of research studies about ecoregional fire regimes can be used to assess burn probability across the nation to identify areas of high risk. This can help government agencies better plan for wildfire hazards. They can also be used as a baseline from which to assess natural fire regimes, which then can be used to abate the threat of fire exclusion and restore fire-adapted ecosystems.

12. Understanding natural processes and the resultant natural patterns they form on a particular region provides essential knowledge about the sustainability of ecosystems, and that collective understanding and knowledge can inspire designs for anthropogenic landscapes that sustain themselves. Designed landscapes should imitate the natural ecosystem patterns of the surrounding ecoregion in which they are embedded. By working with nature's design, people can create landscapes that function sustainably like natural ecosystems.

Appendix A: Air Masses and Frontal Zones

A body of air in which the temperature and moisture are fairly uniform over a large area is known as an **air mass**. The boundary between a given air mass and its neighbor is usually sharply defined. This discontinuity is termed a **front**. In the convergence zone between the tropical polar air masses, winds are variable and high, and accompanied by stormy weather. This zone is called the **polar front**. A large number of the Earth's cyclonic storms are generated here. The properties of an air mass are derived partly from the regions over which it passes. Those land or ocean surfaces that strongly impress their characteristics on the overlying air masses are called **source regions**. Air masses are classified according to their latitudinal position (which determines thermal properties), and underlying surface, whether continent or ocean (determining moisture content). They are summarized in Table 1 and illustrated in Fig. 1.

R.G. Bailey, *Ecoregions*, DOI 10.1007/978-1-4939-0524-9, © Springer Science+Media, LLC 2014

Table 1 Properties of air masses[a]

Major group	Subgroup	Source region	Properties of source
Polar (including arctic)	Polar continental (cP)	Arctic Basin; northern Eurasia and northern North America; Antarctica	Cold, dry, very stable
	Polar maritime (mP)	Oceans poleward of 40 or 50°	Cool, moist, unstable
Tropical (including equatorial)	Tropical continental (cT)	Low-latitude deserts, especially Sahara and Australian Deserts	Hot, very dry, stable
	Tropical maritime (mT)	Oceans of tropics and subtropics instability toward west side of ocean	Warm, moist, greater

[a]From Trewartha et al. (1967)

Fig. 1 Source regions of air masses in relation to the polar front and the intertropical convergence zone (ITC). From *Elements of Physical Geography*, 4th ed., by Arthur N. Strahler and Alan H. Strahler, p. 125. Copyright (c) 1989 by John Wiley & Sons, Inc.; reproduced with permission

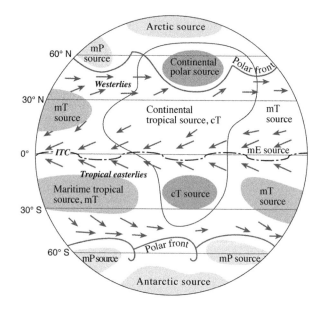

Appendix B: Common and Scientific Names

Plants

Acacia	*Acacia*
Ash	*Fraxinus*
Aspen, quaking	*Populus tremuloides*
Basswood (linden)	*Tilia*
Beech, southern	*Nothofagus*
Birch	*Betula*
Blackeyed Susan	*Rudbeckia hirta*
Bluestem, big	*Andropogon gerardii*
Bluestem, little	*Schizachyrium scoparium*
Buffalograss	*Bouteloua dactyloides*
Cactus apple (prickly-pear)	*Opuntina engelmanni*
Cactus, saguaro	*Carnegiea gigantea*
Cedro espino	*Pachira quinata*
Cheatgrass	*Bromus tectorum*
Chestnut (American)	*Castanea dentata*
Creosote bush	*Larrea tridentata*
Douglas-fir	*Pseudotsuga menziesii*
Eucalyptus (gum)	*Eucalyptus*
Elm	*Ulmus*
Fir	*Abies*
Hemlock	*Tsuga*
Hemlock, western	*Tsuga heterophylla*
Hickory	*Carya*
Hornbeam, American	*Carpinus caroliniana*
Juniper	*Juniperus*
Kauri	*Agathis australis*
Larch (tamarack)	*Larix*
Laurel	*Kalmia*
Locoweed	*Oxytropis*
Magnolia	*Magnolia*
Maple	*Acer*
Maple, sugar	*Acer saccharum*
Mesquite	*Prosopis*
Oak	*Quercus*
Oak, cork	*Quercus suber*
Ocotillo	*Fouquieria splendens*
Pine	*Pinus*
Pine, lodgepole	*Pinus contorta*
Pine, ponderosa	*Pinus ponderosa*
Plum pine	*Podocarpus*
Popular (cottonwood)	*Populus*
Pinyon	*Pinus edulis*
Redcedar, western	*Thuja plicata*
Redwood	*Sequoia sempervirens*
Sagebrush	*Artemisia*
Smoketree	*Psorothamnus spinosus*
Spruce	*Picea*
Spruce, white	*Picea glauca*
Sunflower, common	*Helianthus annuus*
Tamarisk (salt-cedar)	*Tamarix gallica*
Tuliptree	*Liriodendron tulipifera*
Walnut	*Juglans*
Willow	*Salix*

Animals

Antelope (see pronghorn)	
Badger, American	*Taxidea taxus*
Bear	Ursidae
Bear, polar	*Ursus maritimus*
Beaver, American	*Castor canadensis*
Bison, American	*Bison bison*
Bongo	*Tragelaphus eurycerus*
Buffalo, African (Cape buffalo)	*Syncerus caffer*
Camel	*Camelus*
Caribou (reindeer)	*Rangifer tarandus*
Chipmunk	*Tamias*
Chipmunk, alpine	*Tamias alpinus*
Crocodile	*Crocodylus*
Deer	Cervidae

(continued)

R.G. Bailey, *Ecoregions*, DOI 10.1007/978-1-4939-0524-9, © Springer Science+Media, LLC 2014

Elephant, African bush	*Loxodonta africana*
Elephant seal, southern (sea elephant)	*Mirounga leonina*
Elk (red deer)	*Cervus elaphus*
Elk, American (elk, wapiti)	*Cervus elaphus canadensis*
Ermine	*Mustela erminea*
Fox	*Vulpes* and *Alopex*
Gnu (see wildebeest)	
Goat, mountain	*Oreamnos americanus*
Guanaco	*Lama glama guanicoe*
Hamster	Muridae: Cricetidae
Hare, arctic	*Lepus arcticus*
Hippopotamus, common	*Hippopotamus amphibius*
Hog sucker, northern	*Hypentelium nigricans*
Ibex	*Capra ibex*
Kiwi	*Apteryx*
Krill	*Euphausia superba*
Lemming	*Dicrostonyx* and *Lemmus*
Lemur	Lemuridae
Lion, mountain (see puma)	
Marmot, yellow-bellied	*Marmota flaviventris*
Marten, American	*Martes americana*
Mink, American	*Neovison vison*
Moose, American	*Alces americanus*
Moose, Eurasian (elk)	*Alces alces*
Muskox	*Ovibos moschatus*
Okapi	*Okapia johnstoni*
Panther (see puma)	
Penguin	Spheniscidae
Pika, American	*Ochotona princeps*
Plover	*Pluvialis* and *Charadrius*
Prairie dog	*Cynomys*
Pronghorn (antelope)	*Antilocapra americana*
Puma (mountain lion, panther)	*Puma concolor*
Reindeer (see caribou)	
Sable	*Martes zibellina*
Sea elephant (see elephant seal)	
Turkey, wild	*Meleagris gallopavo*
Walrus	*Odobenus rosmarus*
Whale, blue	*Balaenoptera musculus*
Whale, fin (finback)	*Balaenoptera physalus*
Wildebeest, blue (gnu)	*Connochaetes taurinus*
Wolf, gray	*Canis lupus*
Zebra, Burchell's	*Equus burchelli*

Appendix C: Conversion Factors

For readers who wish to convert measurements from the metric system of units to the inch–pound–Fahrenheit system, conversion factors are listed below.

Multiply	By	To obtain
Millimeters	0.039	Inches
Centimeters	0.394	Inches
Meters	3.281	Feet
Kilometers	0.621	Miles
Square meters	10.764	Square feet
Square kilometers	0.386	Square miles
Hectares	2.471	Acres
Centigrade	1.8 + 32	Fahrenheit

R.G. Bailey, *Ecoregions*, DOI 10.1007/978-1-4939-0524-9, © Springer Science+Media, LLC 2014

Appendix D: Comparison of Ecoregion and Related Approaches

This appendix provides relevant information about the currently available climatic, biotic, and ecological regionalization maps that cover the whole globe or large parts of it. An account of the relationship between climate and vegetation, and the theory behind defining ecological zones, or ecoregions, is given in the main text of this book. Olstad (2012) reviews the concepts and uses of ecoregions and the process of ecoregionalization to illustrate the traditions and philosophies within the discipline of geography. Ecoregions for environmental management is the subject of a special issue of *Environmental Management* (Loveland 2004). The ecoregions collection of the *Encyclopedia of Earth* (Cleveland 2011) holds the results from several of the most widely used systems to delineate ecoregions.

The global maps described below primarily define climatic ecological zones. Some of them, such as the World Wildlife Fund (WWF) maps, emphasize differences in biogeography or species origins. Some regional and national maps also emphasize the biogeographical/phylogenetic aspect. Other maps are empirical in that they are based on cluster analysis of data or overlay of thematic maps to identify relative homogeneous areas. But before we review the various ecological regionalization maps, we look first at other related approaches.

Ecoregions Versus Other Land Divisions

Contrasting other land-division categories such as *physiographic regions* and biotic areas (also called "biotic provinces" or "bioregions"), ecoregions are based on both biotic and abiotic features. A geologist might look at a given area in terms only of geologic formations and structures. In fact, a geologist produced one of the best-known physiographic maps of the United States with this perspective (Fenneman 1928). Where major physiographic discontinuities occur—where mountains meet plains, or where *igneous rock* ends and sedimentary rock begins—the boundaries often coincide with changes in the *biota*; quite simply: changes in the land can correlate to changes in flora and its associated fauna. In areas of little relief, such as the Great Plains, little or no correlation exists between the geologist's concept of physiography and the biologist's concept of ecology.

A biologist (c.f. Dice 1943) might examine the same area as the geologist but in terms of the biota's spatial patterns. Large, relatively homogeneous units of biota at the regional scale are known as *biomes* (Clements and Shelford 1939; Brown et al. 1998). Heinrich Walter (1984) refers to these as zonobiomes because they are based on large climatic zones. Subdivisions of biomes have been mapped by Miklos Udvardy (1975) after work started by Raymond Dasmann (1972); these subdivisions are called "biogeographic provinces." However, biota constantly changes due to disturbance and *succession*. For example, either fire or timber harvesting may destroy a forest causing flora-specific fauna either to emigrate or to perish; either outcome produces a profound but temporary absence. As the succession process restores the forest to predisturbance conditions, most if not all of the fauna will repopulate the forest, though it will do so at varying rates according to species. Additionally but quite separately, the geographic distribution of animal species or communities may change due to hunting, a circumstance usually independent of habitat loss. This reality needs to be understood in a multicultural context because not all cultures limit or otherwise regulate hunting in a manner that sustains game species populations.

In contrast to systems that accommodate human influence by recognizing "anthropogenic biomes" (e.g., Ellis and Ramankutty 2008), we need to recognize naturally occurring ecosystem boundaries so that we can more meaningfully map them by screening out the effects of disturbance and succession. Mapping based on present biota characteristics and combinations does not allow for such screening; but screening allows us to recognize, to compare, and to work with ecosystems regardless of land use or disturbance. The potential of any system makes it possible to understand and manage it wisely. One way to illuminate this potential is through the concept of the *climax*. This concept, developed largely by Frederick Clements (1916), recognized that vegetation develops through a series of stages until the whole region is clothed with a uniform plant and animal community. The final stage is determined solely by the climate, and is known as *climatic climax* vegetation. Although theoretically possible, variations in local environments, particularly in soil parent material, prevent large-scale homogeneity so that a region is more likely to have several or many climax vegetation types. This realization led to the polyclimax concept (Tansley 1935), which recognizes that the climax vegetation of a region consists of not just one type but a mosaic of climaxes controlled primarily by soil conditions. These are the *edaphic climaxes*.

To look at any given area regarding the ecological consequences of human activities, we must understand that area's full range of features as understood by geologists, biologists, and others. To the point: we must look at that area from the standpoint of its status as an ecosystem, and not just locally but also much more expansively including the larger land area of which it is a part.

Regarding the delimitation of geographic land units and the purpose driving this endeavor—to create a system for an ecological division of the

world—the term "ecoregion" is comparable to those regions referred to by other authors. Specific examples would include, among others, the following seminal contributions:

- Major natural regions (Herbertson 1905)
- Landscape belts (Passarge 1929)
- Habitat regions (James 1959 et seq.)
- Landscape zone (Isachenko 1973)
- Terrestrial landscapes (Biasutti 1962)
- Morphoclimatic zones (Tricart and Cailleux 1972)

Ecoregions differ from these particular divisional units explicitly because they are based on the distribution of ecosystems. However, the ecoregion concept is much older than it might seem. The ancient Greeks recognized such a concept; and in the eighteenth century, Baron Alexander von Humboldt provided an outline that described latitude zonality and elevation zonality of the plant and animal world in relation to climate (Jackson 2009). The significant work of Dokuchaev (1899) developed the theory of integrated concepts. He explained that extensive areas (zones) share many natural conditions and features in common, but he further explained that these change markedly in passing from one zone to another. One could consider the efforts of C. Hart Merriam (1898) to define *life zones* of the United States as one approach to delineating regions. He described seven transcontinental belts, or life zones, based on associations of plants and animals. His work asserted that these natural zones were suitable to certain varieties of crops; and based on this work, the United States Department of Agriculture later developed the Plant Hardiness Zone Map that divides the country into 11 zones and shows whether a crop or garden plant will survive the average winter (Cathey 1990).

Each ecoregion is typically characterized by a single climax, but two or more climaxes may be represented within a single ecoregion. This often happens on mountains where each elevation zone may have a different climax.

The concept of "ecoregion" differs from that of "biome," for a biome is coincident with its climaxes. Every area having the same climax, however far detached from the main area of that climax, seems to belong to the same biome. An ecoregion, on the contrary, is never discontinuous (except on marine islands), though ecological communities that share similar characteristics may exist in disjunct parts of the world.

Each ecoregion comprises both the climax communities and all the successional stages within its geographic area, and it thus includes the freshwater communities. It does not, however, include the marine communities that may lie adjacent to its shores. These communities belong to the marine ecoregions, which are discussed elsewhere in this volume.

Overviews of terrestrial biomes are common in ecology and physical geography textbooks. Detailed coverage of both terrestrial and marine biomes may be found among the series of books in the Greenwood Guides to Biomes of the World (see Woodward 2009).

Other Ecoregional Classification Systems

Global Mapping

The WWF developed an ecoregion classification system to assess the status of the world's wildlife and conserve the most biologically valuable ecoregions (Olson et al. 2001). The system is well established within the organization's structure and a world ecoregions map has recently been published in collaboration with the National Geographic Society. It was recently distributed to all the US schools. Others have defined ecoregions as areas of ecological potential based on combinations of biophysical parameters such as climate and topography. However, WWF ecoregions emphasize the distribution of distinct biotas, which often do not correspond to zones of ecological potential for reasons such as historical human activities, and chance events, or the complex differences regarding how living communities respond to often subtle environmental conditions. Such mapping relies heavily on expert judgment of species richness and endemism, unique higher taxa, unusual ecological or evolutionary phenomena, and rarity of major habitat types. The WWF-ecoregion framework has been compared unfavorably to the Bailey ecoregion system by Jepson and Whittaker (2001).

Building on the work of the WWF, Abell et al. (2008) developed a regionalization of freshwater ecoregions of the world. They are large areas encompassing one or more freshwater systems with a distinct assemblage of natural freshwater communities and species. They based their ecoregion delineations on qualitative assessments of similarity/dissimilarity using hydrological regions or watershed boundaries and expert input.

The Nature Conservancy (TNC) developed a global regionalization of coastal and shelf areas called marine ecoregions (Spalding et al. 2007). The approach relies on literature review and consultation with experts. They also developed a spatial layer for the terrestrial ecoregions of the world. This layer is based on WWF's ecoregions outside the United States and loosely based on Bailey's ecoregions within the United States. This work parallels another ongoing effort to derive a similar classification of the world's *pelagic* oceans (Spalding et al. 2012).

The Foreign Agricultural Organization (FAO) (Simons 2001) of the United Nations developed a global ecological zoning system based on ecoregional concepts to collect information for the Forest Resource Assessment 2000. The FAO system follows Bailey (1989, 1995, 1998) in using Köppen's climatic classification as a basis for the delineation of zones. Mapping was carried out using potential vegetation maps to define boundaries of climatic zones at three hierarchical levels. The system of global ecological zones has recently been revised and updated (Foreign Agricultural Organization 2012).

The life zone system of Holdridge (1947) uses annual precipitation and "biotemperature" (average days/year without temperatures <0 nor $>30°$ C and potential evapotranspiration ratio, combining precipitation and biotemperature in an index). This life zone system differentiates 38 life zone types. As average data are used, the regions characterized by seasonal pattern are not depicted in an appropriate way (Schwabe and Kratochwil 2011). The Holdridge system has been reviewed in relation to several ecosystem mapping schemes, including Bailey, by Lugo et al. (1999). Harris (1973) commented on the inadequacy of the system to predict soil patterns in Costa Rica.

Walter and Box (1976) developed an ecological classification of the world's climates. It is based on the climate-diagram patterns of Walter, in which temperature and precipitation are plotted on the ratio 1:2 to show periods of aridity and humidity. Also taken into account are number of frost-free days and other extremes that influence vegetation patterns. Their treatment recognizes nine different major zonobiomes plus modifiers that can be added to distinguish particularly dry, cold, or wet conditions. In addition, major variants within zonobiomes are introduced: pedobiomes characterized by extreme edaphic conditions that cause azonal vegetation; orobiomes that involve mountain ranges with vertical climate zonation and elevation belts of vegetation. Zonoecotones, transition zones between individual zonobiomes, are also distinguished. Climate zones are determined by drawing lines around climate stations with similar-appearing climate diagrams as opposed to applying climate formulas.

The boundaries of ecozones developed by Schultz (1995 et seq.) follow the subdivision of the earth into climatic zones established by Troll and Paffen (1964) in their map, *Seasonal Climates of the Earth*. Some geographical regions are difficult to integrate into one ecozone; these are indicated as transitional regions. A few of the ecozones are subdivided into comparatively independent subregions, such as Polar/Subpolar zone, which is divided into tundra and frost debris zone, and ice deserts. The ecozone map elaborated by Schultz has overlaps to a great extent with the Walter and Box zonobiome map.

A global map showing the degree of human-induced modification of ecosystems was developed at Moscow State University (Milanova and Kushlin 1993). It reflects the degree of transformation of present-day landscapes (ecosystems). This map was prepared using existing maps also authored by Moscow State University entitled "Geographical belts and zonal types of landscapes of the world" and "Land use types of the world." Each map unit is given a letter indicating degree of alteration—whether it is virtually undisturbed, with moderate interference (e.g., secondary vegetation), with strong interference (crop cultivation), or with extremely strong change (e.g., towns). This map is useful in defining the status of ecosystems, which is essential to their conservation.

Regional and National Ecoregion Mapping

TNC (Comer et al. 2003) developed a terrestrial and marine ecoregions map of the United States that was originally based on boundaries by Bailey (1995). These boundaries have been extensively modified by TNC's ecoregional planning teams; written justification for each modification is available through TNC's Ecoregional Planning Office.

Roger Sayre et al. (2009) of the U.S. Geological Survey in cooperation with TNC recently produced a map of mesoscale terrestrial ecosystems for the conterminous United States. The mapping system was recently extended to Africa (Sayre et al. 2013). Ecosystems were identified deductively from the top down by combining data layers for biogeography, bioclimate, surficial materials, lithology, land-surface forms, and topographic moisture potential. The methodology was first developed from South America using concept advanced by Huggett (1995) and Bailey (1996).

The U.S. Environmental Protection Agency (EPA) has adopted an empirical ecoregion system developed by Omernik (1987). Boundaries are determined from overlaying maps of ecosystem components: land-surface form, land use, potential natural vegetation, and soils. Ecoregion boundaries are located by subjectively determined coincidence of the component map boundaries. Factors weighted most heavily depend on location and scale. For example, precipitation may be more important in one area while elevation is the most important factor in another area. Ecoregions are subdivided into different levels with Level I the coarsest scale and Level IV the finest scale. Map boundary lines may vary depending on what level is used. The EPA approach has been applied to a number of states, including Alaska (Gallant et al. 1995), to refine the national system. Nowacki et al. (2002) combined the Bailey and Omernik approach to ecoregion mapping in Alaska.

In Canada, the Ecoregion Working Group (1989) developed a map of ecoclimatic regions. Also, the Commission for Environmental Cooperation (2006) produced a map of North America by combining systems for Canada (Wiken 1986), the United States (Omernik 1987), and Mexico. Similar concepts of ecological regionalization have evolved in both Canada and the United States (Bailey et al. 1985).

The Sierra Club (Elder 1994) has created a "critical ecoregions" program designed to protect and restore 21 regional ecosystems in the United States and Canada. They recently targeted ten of them to create climate-resilient habitats where plants, animals, and humans are able to survive on a warmer planet. The basis for the regions is not specified.

Maps of individual continents have been produced following the WWF system, e.g., the *Interim Biogeographic Regionalization of Australia* (IBRA) (Commonwealth of Australia 2012). The *Digital Map of European Ecological Regions* (European Environment Agency 2002) shows the continent divided into 68 regions. It is based on knowledge of climatic and both topographic and geobotanical data, as well as the opinion of a large team of experts from various European nature-related Institutions and the WWF.

Recently, Blasi et al. (2010) defined and mapped the *Ecoregions of Italy* according to a divisive, "top down," approach, from the global macroclimatic domains and divisions of Bailey (1989) down to more detailed units based on an analysis of potential natural vegetation. Where potential natural vegetation has been altered by · human intervention, a multidiscipline team used

significant variations in physical components of the environment (e.g., climate, physiography, soils, and hydrography) to indirectly delineate ecological boundaries.

A long history of eco-geographic regionalization in China goes back as early as 500 B.C. The modern regionalization work began in the mid-twentieth century and culminated recently with a map of China's ecoregions divided into a hierarchy of ecoregions units (Wu et al. 2003a). The first level unit, temperature zone, is delineated with the main criteria of temperature. The second level unit, humidity region, is based on criteria of water/moisture states. The third level unit, natural region, is divided according to medium-size geomorphologic units. Vegetation types and soils are applied as supplementary criteria. Of the existing ecoregional systems, Bailey's systems for the United States, North America, and the continents are comparatively close to China's system in hierarchical units, mapping procedures, and regions (Wu et al. 2003b). Ecological regionalization is a base for rational management and sustainable utilization of ecosystems and natural resources in China (Fu et al. 2004).

An ecoregion map of Japan has been produced by Chen and Morimoto (2009). The motivation for the ecoregion mapping project was twofold: (1) to characterize Japanese watersheds from an ecoregion perspective so as to provide a framework for nationwide scaled ecosystem management; and (2) to suggest demarcation of new political and administrative regions of Japan in an ecological perspective. Of the map's two levels, Ecoregion I is macroscale based on climate; and Ecoregion II is mesoscale based on major landform and geological classes.

Lastly, ecoregions have been addressed quantitatively. In this approach, the goal of ecoregion delineation is to create regions that are internally homogeneous and distinct from other regions regarding a particular set of variables. Given numerical input data, homogeneity can be defined statistically; and algorithms can be applied to sort and divide observations into statistically homogeneous groups ergo regions. Clusters are defined as homogenous groups of observations, and cluster centroids defined by the mean values are used to describe each group. Proponents of this empirical approach claim that it brings no preconception to the mapping task but that it does emphasize methods of numerical taxonomy in the search for pattern recognition. Clustering algorithms have been used (c.f. Hargrove and Luxmore 1998; Host et al. 1996; Mackey et al. 2008; Snelder et al. 2009; Weigelt et al. 2013) to produce ecological regionalization based on environmental factors believed to control ecosystem patterns. They have been criticized as inefficient (Rowe and Sheard 1981) and some argue that boundaries of ecoregions cannot be derived mathematically from one or more datasets without understanding ecosystem pattern and process (Bailey 2004).

Glossary of Technical Terms

Air mass A large and essentially homogeneous body of air, many thousands of km^2 in area, characterized by uniform temperature and humidity.

Alfisol Soil order consisting of soils of humid and subhumid climates, with high-base status and argillic horizon.

Alkali Salts found in soils, as in some deserts.

Anadromous fisheries Migrating from salt water to spawn in fresh water, such as salmon.

Aridisol Soil order consisting soils of dry climates, with or without argillic horizons, and with accumulations of carbonates or soluble salts.

Arroyo In Southwest USA, steep-sided dry valley, usually inset in alluvium.

Azonal Zonal in a neighboring zone but confined to an extrazonal environment in a given zone; mountains that cut across the lowland ecological zones, or regions.

Biogeographic province Subdivision of a biome, based on animal and plant distribution.

Biogeographical region One of the eight continent or subcontinent-sized areas of the biosphere, each representing evolutionary core areas of related fauna and flora; e.g., the Neotropical of Walace (1876).

Biomass The dry mass of all living materials in an area.

Biome Geographical region classified on the basis of dominant vegetation and main climate; e.g., the temperate biome is the geographical area with a temperate climate and forests composed of mixed deciduous tree species.

Biophysical factor Elements, such as latitude, continental position, and elevation, that cause different climates and associated ecosystems.

Bioregion (Also called *biotic province* and *biotic area*) geographic expanse which corresponds to the distribution of one or more groups of living beings, usually animals; e.g., the Carolinian bioregion is characterized by the tulip tree, the raccoon, and so on.

Biota Plant and animal life of a region.

Black prairie soil (Also called *prairie soil* or Brunizem) acid grassland soil.

Bog A wet area covered by acid peat.

Bolson From a Spanish word, meaning pocket, for a basin of inland drainage.

Boreal forest See *tayga*.

Broad leafed With leaves other than linear in outline; as opposed to needle-leafed or grasslike (graminoid).

Brown forest soil (Also called *gray brown podzolic*) acid soil with dark brown surface layers, rich in humus, grading through lighter colored soil layers to limy parent material; develops under deciduous forest.

Brown soil Alkaline soil having thin brown surface layer that grades downward into a layer where carbonates have accumulated; develops under grasses and shrubs in semiarid environments.

Calcification Accumulation of calcium carbonate in a soil

Carbon sequestration Capture and long-term storage of atmospheric carbon dioxide (CO_2)

Chaparral Sclerophyll scrub and dwarf forest found throughout the coastal mountain ranges and hills of central and southern California.

R.G. Bailey, *Ecoregions*, DOI 10.1007/978-1-4939-0524-9, © Springer Science+Media, LLC 2014

Chernozem Fertile, black or dark brown soil under prairie or grassland with lime layer at some depth between 0.6 and 1.5 m.

Chestnut-brown soil Short-grass soil in subhumid to semiarid climate with dark brown layer at top, which is thinner and browner than in chernozem soils, that grades downward to a layer of lime accumulation.

Climate Generalized statement of the prevailing weather conditions at a given place, based on statistics of a long period of record.

Climate diagram According to Walter (1984): a diagram in which months are on the horizontal axis and extend January to December for the Northern Hemisphere and July to June for the Southern Hemisphere so summer is always in the diagram middle and curves give mean monthly values of temperature in centigrade and rainfall in millimeters; by choosing a scale at which 10 °C corresponds to 20 mm of rainfall, a relatively dry season (rainfall curve lies below the temperature curve) can be depicted.

Climatic climax The relatively stable community that terminates on zonal soils.

Climatic regime Seasonality of temperature and moisture.

Climax Relatively stable state of the vegetation.

Clustering algorithm A step-by-step procedure for grouping sets of objects according to similar characteristics.

Compensating factor Factor, or condition, that overrides other factors to bring about the same result.

Continental shelf The edge of a continent submerged in relatively shallow seas and oceans.

Cyclone Whirling storm characteristic of middle latitudes; any rotating low-pressure air system.

Deciduous Woody plants, or pertaining to woody plants, that seasonally lose all their leaves and become temporarily bare stemmed.

Delta The flat alluvial area at the mouth of some rivers, where the mainstream splits into several distributaries.

Desert Supporting vegetation of plants so widely spaced, or sparse, that enough of the substratum shows through to give the dominant tone to the landscape.

Desertification Degradation of the plant cover and soil as a result of overuse, especially during periods of drought.

Desert soil Shallow, gray soils containing little humus, excessive amounts of calcium carbonate at depths less than 30 cm.

Desert-like savanna Tropical semidesert with scattered low trees or shrubs.

Division As defined for use in this book: a subdivision of a domain determined by isolating areas of definite vegetation affinities that fall within the same regional climate (continents) or areas of similar water temperature, salinity, and currents (oceans).

Doldrums An area near the equator of very ill-defined surface winds associated with the intertropical convergence zone.

Domain As defined for use in this book: groups of ecoregions with related climates (continents) or water masses (oceans).

Dry steppe See *dry savanna*.

Dry savanna or steppe With 6–7 arid months in each year.

Ecoclimatic unit Ecosystem unit based on climate.

Ecological restoration Practice of renewing and restoring degraded, damaged, or destroyed ecosystems and habitats by active human intervention.

Ecoregion (Also called *ecosystem region*) major ecosystem, resulting from large-scale predictable patterns of solar radiation and moisture, which in turn, affect the kinds of local ecosystems and animals and plants found there.

Ecoregional design Outcome of a deliberative and analytical decision-making process for planning that minimizes environmental destructive impacts by integrating itself with the characteristics and processes of the ecoregion.

Ecosystem An area of any size with an association of physical and biological components so organized that a change in any one component will bring about a change in the other components, and in the operation of the whole system.

Edaphic Pertaining to soil.

Edaphic climax Stable community of plants that develops on soils different from those supporting a climatic climax.

Elevational zonation Vertical differentiation of climate, vegetation, and soil based on the effects of elevation change.

Empirical Source of knowledge acquired by means of observation or experimentation.

Epiphyte Organism that lives on the surface of a plant, but does not draw nourishment from it.

Erg A very large area of sand dunes within a desert.

Evergreen Plants, or pertaining to plants, which remain green in parts the year around, either by retaining at least some of their leaves at all times, or by having green stems which carry on photosynthesis.

Exotic river Stream that flows across a region of dry climate and derives its discharge from adjacent uplands where a water surplus exists.

Fjord A deeply glaciated valley in a coastal region

Fire cycle See *fire regime*

Fire regime Character of fire occurrence; e.g., frequent surface fire.

Fire regime condition class A classification of the amount of departure from the natural fire regime.

Forest Open or closed vegetation with the principal layer consisting of trees averaging more than 5 m in height.

Forest Inventory & Analysis (FIA) USDA Forest Service program provides information to assess the nation's forests.

Forest-steppe Intermingling of steppe and groves or strips of trees.

Forest-tundra Intermingling of forest and tundra.

Formative process A set of actions and changes that occur in the landscape through collective geomorphic, climatic, biotic, and cultural activities.

Front Division between two air masses with different origins and different characteristics.

Galeria forest Dense tropical, or prairie, forest living along the banks of a river.

General circulation model (GCM) A class of computer-driven models for forecasting weather, understanding climate, and projecting climate change.

Geostrophic Pertaining to deflective force due to rotation of the earth.

Germplasm Substance of the germ cells by which the hereditary characteristics are believed to be transmitted.

Grassy savanna Savanna in which woody plants are entirely lacking.

Gray brown podzol soil Acid soil under broadleaf deciduous forest; has thin, organic layer over grayish brown, leached layer; layer of deposition is darker brown.

Greenhouse effect Accumulation of heat in the lower atmosphere resulting from the absorption of long-wave radiation from the earth's surface.

Growth and yield model A set of relationships, usually expressed as equations and embodied in a computer program, that provides estimates of future stand development given initial stand conditions and a specified management regime.

Hamada An eroded rock-surface found in deserts.

Histosol Soil order consisting of soils which are organic.

Horse latitudes Subtropical high-pressure belt of the oceans.

Humus Organic material derived, by partial decay, from the organs of dead plants.

Hydrograph Graph showing the rate of flow (discharge) versus time past a specific point in a river or stream.

Igneous rock A type of rock formed by the solidification of magma, either within the earth's crust or at the surface.

Inceptisols Soil order consisting of soils with weakly differentiated horizons showing alteration of parent materials.

Impute Estimate; process of replacing missing data.

Isotherm   Line on a map connecting points of equal temperature.

Intertropical convergence zone   (Commonly abbreviated *ITC*) a broad zone of low pressure, migrating northwards and southwards of the equator with the season, toward which tropical air masses converge.

Intrazonal   Exceptional situations within a zone, e.g., on extreme types of soil that override the climatic effect.

Krummholz   Zone of wind-deformed trees between the montane and alpine zones.

Landscape   *See* landscape mosaic.

Landscape mosaic   As defined for use in this book: a geographic group of site-level ecosystems.

Laterite   A residual soil developing in the tropics, containing concentrations of iron and aluminum hydroxides which stain the soil red.

Laterization   Process of forming laterite.

Latisol   Major soil type associated with humid tropics, and characterized by red, reddish brown or yellow coloring.

Lichen   Combinations of algae and fungi living together symbiotically; typically form tough, leathery coatings or crusts attached to rocks and tree trunks.

Life zone   Temperature-based concept developed by C. Hart Merriam in 1898 as a means of describing areas with similar plant and animal communities; revised in 1947 by Holdridge, who proposed a life zone classification based on indicators of mean annual biotemperature, annual precipitation, and ratio of annual potential evapotranspiration to mean total annual precipitation.

Light tayga   Tayga forest composed of larch and pine or spruce.

Macroclimate   Large climatic zone arranged in a latitudinal band; climate that lies just above the local modifying irregularities of landform and vegetation.

Meadow   Closed herbaceous vegetation, commonly in stands of rather limited extent, or at least not usually applied to extensive grasslands.

Meadow steppe   The steppe component of the forest-steppe zone.

Mollisols   Soil order consisting of soils with a thick, dark-colored, surface-soil horizon, containing substantial amounts of organic matter (humus), and high-base status.

Mixed forest   Forest with both needle-leafed and broad-leafed trees.

Monsoon forest   Drought-deciduous trees.

Non-point source pollution   Refers to both water and air pollution from diffuse sources.

Normalized Differences Vegetation Index (NDVI)   Simple numerical indicator used to analyze remote sensing measurements and assess whether the target being observed contains live green vegetation.

Oceanic polar front   In arctic and antarctic regions, boundary between warm and cold water types, associated with a convergence of surface currents.

Oceanic whirl   Circular movement of air around the subtropical pressure-high zone.

Open woodland   (Also called *steppe forest* and *woodland savanna*) open forest with lower layers also open, having the trees or tufts of vegetation discrete, but averaging *less* than their diameter apart.

Orographic precipitation   Rain, snow, sleet, and so on induced by the forced rise of moist air over a mountain barrier.

Oxisols   Soil order consisting of soils that are mixtures principally of kaolin, hydrated oxides, and quartz.

Paramo   The alpine belt in the wet tropics.

Parkland   Areas where clumps of trees alternate with grassland, but where neither becomes an extensive, uninterrupted stand.

Pelagic   Of the ocean surface, especially as distinguishing from coastal waters.

Permafrost   Permanently frozen soil.

Physiognomy   General overall appearance of vegetation based on plant growth form and maturity stage regardless of floristic composition or dominant species.

Physiography   Landform (including surface geometry and underlying geologic material).

Physiographic region An area of similar geologic structure and topographic relief that has a unified geomorphic history; e.g., the Great Plains of Fenneman (1928).

Piedmont Sequence of landforms along the margins of uplands.

Plankton Small, floating or weakly swimming plants and animals, found in salt and fresh water; primarily microscopic algae and protozoa.

Plant adaptation region Large area based on both ecological (ecoregion) and climatic (hardiness zone) characteristics that can be used to guide collecting and evaluating plant materials for potential adaptations.

Plant formation class A world vegetation type dominated throughout by plants of the same life form.

Plate tectonics The large-scale motions of the earth's lithosphere.

Playa A desert lake existing only temporarily after a rain.

Pleistocene The most recent major ice age. Generally the Pleistocene is considered to have begun approximately two million years ago and to have ended 8 to 10 thousand years ago.

Podzol Soil order consisting of acid soil in which surface soil is strongly leached of bases and clays.

Polar front Boundary lying between cold polar air masses and warm tropical air masses.

Potential natural vegetation Naturally occurring plant cover known to occur in areas undisturbed by human activity and assumed to grow in a disturbed area if human intervention should be removed.

Prairie (Also known as *tall-grass prairie*) grassland characterized by grasses 1 m tall or taller, those grasses growing close together and exposing little or no bare soil, and shrubs conspicuously absent except for isolated site-specific patches.

Predictive model See *growth and yield model*.

Regolith Layer of weathered inorganic and organic debris overlying the surface of the earth.

Salinity A saline quality

Salinization Precipitation of soluble salts within the soil.

Savanna forest The forest component of the savanna.

Savanna Closed grass or other predominantly herbaceous vegetation with scattered or widely spaced woody plants usually including some low trees.

Seed transfer zone Identifiable area, usually with definite topographic bounds, climate, and growing conditions, containing plants with relatively uniform genetic (racial) composition.

Sclerophyll or sclerophyllous Refers to plants with predominantly hard, stiff leaves that are usually evergreen.

Selva An alternative term for tropical rainforest, originally applied to the Amazon Basin.

Semideciduous forest Composed partly of evergreen and partly deciduous species.

Semidesert (Also called *half-desert*) is an area of xerophytic shrubby vegetation with a poorly developed herbaceous lower layer, e. g., sagebrush.

Shrub savanna Closed grass or other predominantly herbaceous vegetation with scattered or widely spaced shrubs.

Shrub A woody plant less than 5 m in height.

Sierozem See desert soil.

Silvicultural practices Generally: the science and art of cultivating and managing forest crops based on silvics, the collective knowledge of both tree and forest ecology.

Site The smallest, or local, ecosystems.

Small-leafed As used here, refers to birch and aspen.

Soil *Great group* third level of classification of soils, defined by similarities in kind, arrangement, and distinctiveness of horizons, as well as close similarities in moisture and temperature regimes, and base status.

Soil orders Those ten soil classes forming the highest category in the classification of soils.

Source region Extensive land or ocean surface over which an air mass derives its characteristics.

Spodosols   Soil order consisting of soils that have accumulations of amorphous materials in subsurface horizons.

Steppe   (*Also called short-grass prairie*) grassland–shrubland mix characterized by grasses less than 1 m tall, those grasses widely spaced exposing much bare soil often grown with lichens, and shrubs present and typically conspicuous though often small and dispersed.

Subtropical high-pressure belts   (Also called *cells* or *zones*) belts of persistent high atmospheric pressure tending east–west and centered at about lat. 30°N and S.

Succession   The replacement of one community of plants and animals by another.

Sustainable design   The process of prescribing compatible land uses and building based on the limits of place, locally as well as regionally.

Taxon   A group of objects; based on the similarity of properties.

Tayga   (Also spelled *taiga*) a parkland or savanna with needle-leafed (usually evergreen) low trees and shrubs; a Russian word referring to the northern virgin forests.

Temperate rainforest   Dense forest, comprising tall trees, growing in areas of very high rainfall, such as the Pacific Northwest of the USA.

Thermoisopleth diagram   Drawing that shows temperature at a station throughout the day for every day of the year.

Thermokarst   The formation of a highly irregular ground surface, as a result of the thawing of masses of ground ice.

Topoclimate   The climate of a very small space; influenced by topography.

Toposequence   A change of a community with topography.

Trade winds   Current of air blowing from the east on the equatorward side of the subtropical high-pressure cells.

Transhumance   The seasonal movement of people and animals to and from fresh pastures.

Tundra   Slow-growing, low formation, mainly closed vegetation of dwarf shrubs, graminoids, and cryptograms, beyond the subpolar or alpine treeline.

Tundra soil   Cold, poorly drained, thin layers of sandy clay and raw humus; without distinctive soil profiles.

Upwelling   Upward motion of cold, nutrient-rich ocean waters, often associated with cool equatorward currents occurring along western continental margins.

Ultisols   Soil order consisting of soils with horizons of clay accumulation and low base supply.

Wadi   In Arabia and the Sahara, dry desert valley.

Westerlies   Winds blowing from the west on the poleward side of the subtropical high-pressure cells.

Wildfire occurrence gradient   Differences in the number of fires in non-urban settings.

Wildland urban interface   Zone of transition between unoccupied land and human development. These lands and communities adjacent to and surrounded by wildlands are at risk of wildfires.

Woodland   Cover of trees whose crowns do not mesh, with the result that branches extend to the ground.

Xerophyte   A plant adapted to an environment characterized by extreme drought.

Yellow forest soil   (Also called *red-yellow podzol*) soils with weakly developed horizons but also have accumulations of sesquioxides of iron and aluminum; transitional between podzols and latosols.

Zonal   Resulting from the average state of the atmosphere; variation in environmental conditions in a north–south direction.

Zonal soil   Well-developed deep soils on moderate surface slopes that are well drained.

Bibliography

On Ecological Divisions of the Oceans

Bogorov VG (1962) Problems of the zonality of the world ocean. In: Harris CD (ed) Soviet geography, accomplishments and tasks. Occasional publication no. 1. American Geographical Society, New York, pp 188–194

Longhurst AR (2006) Ecological geography of the sea, 2nd edn. Academic Press, San Diego, 560 pp

Sherman K, Alexander LM, Gold BD (eds) (1990) Large marine ecosystem: patterns, processes, and yields. American Association for the Advancement of Science, Washington, DC, 242 pp

Terrell TT (1979) Physical regionalization of coastal ecosystems of the United States and its territories. FWS/OBS-78/80. U.S. Fish and Wildlife Service, Washington, DC, 30 pp

Woodward SL (2008) Marine biomes. Greenwood Press, Westport, CT, 212 pp

On the Climatic and Ecological Divisions of the Continents as a Whole or Larger Parts

Anon (1995) World ecoregions, types of natural landscapes. In: Espenshade EB, Hudson JC, Morrison JL (eds) Goode's world atlas (19th edn). Rand McNally, Chicago, pp 22–23. 1:77,000,000

Akin WE (1991) Global patterns: climate, vegetation, and soils. University of Oklahoma Press, Norman, 370 pp

Allee WC, Schmidt KP (1951) Ecological animal geography (based on *Tiergeographie auf oekologische Grundlage* by Richard Hesse), 2nd edn. Wiley, New York, 715 pp

Atwood WW (1940) The physiographic provinces of North America. Ginn, Boston, 536 pp

Austin ME (1965) Land resource regions and major land resource areas of the United States (exclusive of Alaska and Hawaii). Agriculture handbook 296. USDA Soil Conservation Service, Washington, DC, 82 pp. With separate map at 1:7,500,000

Bailey RG (1976) Ecoregions of the United States. USDA Forest Service, Intermountain Region, Ogden, UT. 1:7,500,000; colored

Bailey RG (1983) Delineation of ecosystem regions. Environ Manage 7:365–373

Bailey RG (2002) Ecoregions (Chap 12). In: Orme AR (ed) The physical geography of North America. Oxford University Press, New York, pp 235–245

Bailey RG, Cushwa CT (1981) Ecoregions of North America. FWS/OBS-81/29. U.S. Fish and Wildlife Service, Washington, DC. 1:12,000,000; colored

Barnes BV (1984) Forest ecosystem classification and mapping in Baden-Württemberg, West Germany. In: Bockheim JG (ed) Proceedings, forest land classification: experiences, problems, perspectives, Madison, WI, 18–20 March 1984, pp 49–65

Bashkin VN, Bailey RG (1993) Revision of map of ecoregions of the world (1992–95). Environ Conserv 20:75–76

Bear FE, Pritchard W, Akin WE (1986) Earth: the stuff of life, 2nd edn. University of Oklahoma Press, Norman, 318 pp

Bennett CF (1975) Man and Earth's ecosystems. Wiley, New York, 331 pp

Berg LS (1947) Geograficheskiye zony Sovetskogo Soyuza (Geographical zones of the Soviet Union), vol 1, 3rd edn. Geografgiz, Moscow

Billings WD (1964) Plants and the ecosystem. Wadsworth, Belmont, CA, 154 pp

Birot P (1970) Les regions naturelles du globe. Masson, Paris, 380 pp

Bockheim JG (2005) Soil endemism and its relation to soil formation theory. Geoderma 129:109–124

Bourne R (1931) Regional survey and its relation to stocktaking of the agricultural and forest resources of the British Empire. Oxford Forestry memoirs 13. Clarendon Press, Oxford, 169 pp

Bowman I (1911) Forest physiography, physiography of the U.S. and principal soils in relation to forestry. Wiley, New York, 759 pp

Brazilevich NI, Rodin LY, Rozov NN (1971) Geographical aspects of biological productivity. Sov Geogr 12:293–317

Breckle S-W (ed) (2002) Walter's vegetation of the Earth: the ecological systems of the geo-biosphere, 4th edn. Springer, Berlin, 527 pp

Breymeyer AI (1981) Monitoring of the functioning of ecosystems. Environ Monit Assess 1:175–183

Budyko MI (1974) Climate and life (English edition by D. H. Miller). Academic Press, New York, 508 pp

Dansereau P (1957) Biogeography–an ecological perspective. Ronald Press, New York, 394 pp

de Laubenfels DJ (1970) A geography of plants and animals. WM. C. Brown, Dubuque, IA, 133 pp

de Laubenfels DJ (1975) Mapping the world's vegetation: regionalization of formations and flora. Syracuse University Press, Syracuse, 246 pp

Delvaux J, Galoux A (1962) Les territoires écologiques du sud-est belge. Centre d'Ecologie generale. Travaux hors _erie, 311 pp

Denton SR, Barnes BV (1988) An ecological climatic classification of Michigan: a quantitative approach. For Sci 34(1):119–138

Eyre SR (1963) Vegetation and soils: a world picture. Aldine Publishing, Chicago, 324 pp

FAO/UNESCO (1971–1978) FAO/UNESCO soil map of the world 1:5 million. North America, South America, Mexico and Central America, Europe, Africa, South Asia, North and Central Asia, Australia. UNESCO, Paris

Forman RTT (1995) Land mosaics: the ecology of landscapes and regions. Cambridge University Press, Cambridge, 632 pp

Forman RTT, Godron M (1986) Landscape ecology. Wiley, New York, 619 pp

Garner HF (1974) The origins of landscapes: a synthesis of geomorphology. Oxford University Press, New York, 734 pp

Gaussen H (1954) Théorie et classification des climats et microclimats. 8me Congr. Internat. Bot. Paris, Sect. 7 et 3, pp 125–130

Geiger R (1965) The climate near the ground (trans.). Harvard University Press, Cambridge, MA, 611 pp

Gersmehl P, Napton D, Luther J (1982) The spatial transferability of resource interpretations. In: Braun TB (ed) Proceedings, national in-place resource inventories workshop, University of Maine, Orono, 9–14 Aug 1981. Society of American Foresters, Washington, DC, pp 402–405

Gleason HA, Cronquist A (1964) Natural geography of plants. Columbia University Press, New York, 420 pp

Goudie A (1993) The nature of the environment, 3rd edn. Blackwell, Oxford, 397 pp

Grigor'yev AA (1961) The heat and moisture regime and geographic zonality. Sov Geogr Rev Transl 2:3–16

Günther M (1955) Untersuchungen über das Ertragsvermögen der Haupt-holzarten in Bereich verschiederner des württenbergischen Necharlandes. Mitt Vereins f forstl Standortsk u Forstpflz 4:5–31

Haggett P (1972) Geography: a modern synthesis. Harper & Row, New York, 483 pp

Hammond EH (1954) Small-scale continental landform maps. Ann Assoc Am Geogr 44:33–42

Hare T (1994) Habitats: 14 gatefold panoramas of the world's ecological zones. Macmillan, New York, 143 pp

Hills GA (1960) Comparison of forest ecosystems (vegetation and soil) in different climatic zones. Silva Fennica 105:33–39

Hole FD (1978) An approach to landscape analysis with emphasis on soils. Geoderma 21:1–23

Hole FD, Campbell JB (1985) Soil landscape analysis. Rowman & Allanheld, Totowa, NJ, 196 pp

Hopkins AD (1938) Bioclimatics: a science of life and climate relations. Miscellaneous publication no. 280. U.S. Department of Agriculture, Washington, DC, 188 pp

Hou XY (1988) Physical ecoregion of China and mega-agricultural development. Bull Chin Acad Sci 1:28–37; 2:137–152

Howard JA, Mitchell CW (1985) Phytogeomorphology. Wiley, New York, 222 pp

Joerg WLG (1914) The subdivision of North America into natural regions: a preliminary inquiry. Ann Assoc Am Geogr 4:55–83

Krajina VJ (1965) Biogeoclimatic zones and classification of British Columbia. In: Krajina VJ (ed) Ecology of western North America. University of British Columbia Press, Vancouver, BC, pp 1–17

Küchler AW (1974) Boundaries on vegetation maps. In: Tüxen R (ed) Tatsachen und problem dergrenzen in der vegetation. Verlag von J. Cramer, Lehre, Germany, pp 415–427

Kul'batskaya IY (editor-in-chief) (1988) Geograficheskiye poyasa i zonal'nyye tipy landshaftov Mira (Geographic belts and zonal types of landscapes of the world). USSR Academy of Sciences and Main Administration of Geodesy and Cartography. USSR, Moscow (in Russian). 1:15,000,000

Leser H (1976) Landscaftsökologie. Eugen Ulmer, Stuttgart, 432 pp

Lewis GM (1966) Regional ideas and reality in the Cis-Rocky Mountain west. Trans Inst Br Geogr 38:135–150

Lomolino MV, Riddle BR, Whittaker RJ, Brown JH (2010) Biogeography, 4th edn. Sinauer, Sunderland, MA, 878 pp

MacArthur RH (1972) Geographical ecology: patterns in the distribution of species. Princeton University Press, Princeton, NJ, 269 pp

Masing V (1997) Major subdivisions of the biota of the world: some general problems in biogeography. Bot Lith Suppl 1:5–14

Mather JR, Sdasyuk GV (eds) (1991) Global change: geographical approaches. University of Arizona Press, Tucson, 289 pp

McHarg IL (1969) Design with nature. American Museum of Natural History by The Natural History Press, Garden City, NY, 197 pp

Mil'kov FN (1979) The contrastivity principle in landscape geography. Sov Geogr 20:31–40

Miller DH (1978) The factor of scale: ecosystem, landscape mosaic, and region. In: Hammond KA, Macinko G, Fairchild WB (eds) Sourcebook on the environment. University of Chicago Press, Chicago, pp 63–88

Müller P (1974) Aspects of zoogeography. Dr. W. Junk, The Hague, 208 pp

Nielson RP (1987) Biotic regionalization and climatic controls in western North America. Vegetatio 70:135–147

Noss RF (1983) A regional landscape approach to maintaining diversity. Bioscience 33:700–706

Olson JS, Watts JS (1982) Major world ecosystem complexes. In: Carbon in live vegetation of major world ecosystems. ORNL-5862. Oak Ridge National Laboratory, Oak Ridge, TN. 1:30,000,000

Oosting HJ (1956) The study of plant communities, 2nd edn. W.H. Freeman, San Francisco, 440 pp

Orme AR (ed) (2002) The physical geography of North America. Oxford University Press, New York, 551 pp

Pojar J, Klinka K, Meidinger DV (1987) Biogeoclimatic ecosystem classification in British Columbia. For Ecol Manage 22:119–154

Rowe JS (1962) The geographic ecosystem. Ecology 43:575–576

Shantz HL, Marbut CF (1923) The vegetation and soils of Africa. Research series 13. American Geographical Society, New York, 263 pp

Shelford VE (1963) The ecology of North America. University of Illinois Press, Urbana, 609 pp

Smith RL (1996) Ecology and field biology, 5th edn. HarperCollins, New York, 733 pp

Stephenson NL (1990) Climatic control of vegetation distribution: the role of the water balance. Am Nat 135:649–670

Sukachev V, Dylis N (1964) Fundamentals of forest biogeocoenology (trans. from Russian by J.M. Maclennan). Oliver & Bond, London, 672 pp

Swanson FJ, Kratz TK, Caine N, Woodmansee RG (1988) Landform effects on ecosystem patterns and processes. Bioscience 38:92–98

Troll C (1971) Landscape ecology (geoecology) and biogeocenology—a terminology study. Geoforum 8:43–46

Tukhanen S (1986) Delimitation of climatic-phytogeographical regions at the high-latitude area. Nordia 20:105–112

Veatch JO (1930) Natural geographic divisions of land. Mich Acad Sci Arts Lett 19:417–427

Volubief VP (1953) Soils and climate. Azerbaijan Academy of Science, Baku, 319 pp

Walter H (1977) Vegetationszonen und klima: die ökologische gliederung der biogeosphäre. Eugen Ulmer Verlag, Stuttgart, Germany, 309 pp

Walter H, Breckle SW (1985) Ecological systems of the geobiosphere, vol 1: Ecological principles in global perspective (trans. from German by S. Gruber). Springer, Berlin, 242 pp

Whittaker RH (1975) Communities and ecosystems, 2nd edn. MacMillan, New York, 387 pp

Woodward SL (2003) Biomes of Earth: terrestrial, aquatic, and human-dominated. Greenwood Press, Westport, CT, 456 pp

Yoshino MM (1975) Climate in a small area: an introduction to local meteorology. University of Tokyo Press, Tokyo, 549 pp

On Regional Systems: Polar

Hare FK (1950) Climate and zonal divisions of the boreal forest formation in eastern Canada. Geogr Rev 40:615–635

Hare FK, Ritchie JC (1972) The boreal bioclimates. Geogr Rev 62:333–365

Ives JD, Barry RG (eds) (1974) Arctic and alpine environments. Methuen, London, 999 pp

Pielke RA, Vidale PL (1995) The boreal forest and the polar front. J Geophys Res 100:25,755–25,758

Polunin N (1951) The real Arctic: suggestions for its delimitation, subdivision and characterization. J Ecol 39:308–315

Quinn JA (2008) Arctic and alpine biomes. Greenwood Press, Westport, CT, 218 pp

Schlutz J (1995) See Chapters 3.1 and 3.2

Shear JA (1964) The polar marine climate. Ann Assoc Am Geogr 54:310–317

Silver KC, Carroll M (2013) A comparative review of North American tundra delineations. ISPRS Int J Geo-Inf 2:324–348

Tricart J (1970) Geomorphology of cold environments (trans. from French by Edward Watson). Macmillan, London, 320 pp

Humid Temperate

Albert DA (1995) Regional landscape ecosystems of Michigan, Minnesota, and Wisconsin: a working map and classification. U.S. Forest Service General Technical Report NC-178. North Central Forest Experiment Station, St. Paul, MN, 250 pp

Albert DA, Denton SR, Barnes BV (1986) Regional landscape ecosystems of Michigan. University of Michigan, Ann Arbor, 32 pp With separate map at 1:1,000,000

Allen HD (2001) Mediterranean ecogeography. Pearson Education, Harlow, England, 263 pp

Borchert JF (1950) The climate of the central North American grassland. Ann Assoc Am Geogr 40:1–39

Braun EL (1950) Deciduous forests of eastern North America (reprinted 1964). Hafner, New York, 596 pp

Di Castri F, Goodall DW, Specht RL (eds) (1981) Mediterranean-type shrublands. Ecosystems of the World 11. Amsterdam, Elsevier, 643 pp

Dix RL, Smeins FE (1967) The prairie, meadow, and marsh vegetation of Nelson County, North Dakota. Can J Bota 45:21–58

Kuennecke BH (2008) Temperate forest biomes. Greenwood Press, Westport, CT, 193 pp

Orme AT, Bailey RG (1971) Vegetation and channel geometry in Monroe Canyon, southern California. Yearb Assoc Pac Coast Geogr 33:65–82

Ovington JD (ed) (1983) Temperate broad-leaved evergreen forests. Ecosystems of the World 10. Elsevier, Amsterdam, 241 pp

Röhrig E, Ulrich B (eds) (1991) Temperate deciduous forests. Ecosystems of the world 7. Elsevier, Amsterdam, 635 pp

Schultz J (1995) See Chapters 3.3, 3.6, and 3.8

Woodward SL (2008b) Grassland biomes, Chap 2. Greenwood Press, Westport, CT, 148 pp

Dry

Coupland RT (ed) (1992/1993) Natural grasslands. Ecosystems of the world 8A and 8B. Elsevier, Amsterdam, p 469, p 556

Evenari M, Noy-meir I, Goodall DW (eds) (1985/1986) Hot deserts and arid shrublands. Ecosystems of the world 12A and B. Elsevier, Amsterdam, p 365, p 451

Goudie AS, Wilkinson JC (1977) The warm desert environment. Cambridge University Press, Cambridge, 88 pp

Hunt CB (1966) Plant ecology of Death Valley, California. Professional paper 509. U.S. Geological Survey, Washington, DC, 68 pp

Quinn JA (2009) Desert biomes. Greenwood Press, Westport, CT, 226 pp

Schultz J (1995) See Chapters 3.4 and 3.5

Shreve F (1942) The desert vegetation of North America. Bot Rev 8:195–246

UNESCO (1977) Map of the world distribution of arid regions. MAB technical notes 7. United Nations Education, Scientific and Cultural Organization, Paris, 54 pp With separate map at 1:25,000,000

West NE (ed) (1983) Temperate deserts and semi-deserts. Ecosystems of the world 5. Elsevier, Amsterdam, 522 pp

Humid Tropical

Beard JS (1955) The classification of tropical American vegetation types. Ecology 36:89–100

Bourliere F (ed) (1983) Tropical savannas. Ecosystems of the world 13. Elsevier, Amsterdam, 730 pp

Cole MM (1960) Cerrado, caatinga and pantanal: the distribution and origin of the savanna vegetation of Brazil. Geogr J 126:168–179

Fosberg FR, Garnier BJ, Küchler AW (1961) Delimitation of the humid tropics. Geogr Rev 51:333–347. With separate map at 1:60,000,000

Golley FB (ed) (1983) Tropical rain forest ecosystems. Ecosystems of the world 14A. Elsevier, Amsterdam, 381 pp

Holzman BA (2008) Tropical forest biomes. Greenwood Press, Westport, CT, 242 pp

le Houerou HN, Popov GF (1981) An eco-climatic classification of intertropical Africa. FAO technical paper 31. FAO, Rome, 40 pp

Ruhe RV (1960) Elements of the soil landscape. In: Seventh international congress of soil science, vol 23, Adaliele, pp 165–170

Schlutz J (1995) See Chapters 3.7 and 3.9

Tosi JS (1964) Climatic control of terrestrial ecosystems: a report on the Holdridge model. Econ Geogr 40:173–181

Tricart J (1972) The landforms of the humid tropics, forests, and savannas (trans. from French by Conrad J. Kiewiet de Jonge). Longman, London, 306 pp

Woodward SL (2008) See Chapter 3

Mountains

Barry RG (1992) Mountain weather and climate, 2nd edn. Routledge, London, 402 pp

Daubenmire R (1943) Vegetation zonation in the Rocky Mountains. Bot Rev 9:325–393

Ives JD, Messerli M (1997) Mountains of the world: a global priority. In: Messerli M, Ives JD (eds) Mountains of the world: a global priority. Parthenon, New York, pp 1–15

Merriam CH (1890) Results of a biological survey of the San Francisco Mountain region and desert of the Little Colorado, Arizona. North American Fauna, No. 3, pp 1–136

Parish R (2002) Mountain environments. Pearson Education, Harlow, England, 348 pp

Pfister RD, Arno SF (1980) Classifying forest habitat types based on potential climax vegetation. For Sci 26:52–70

Quinn JA (2008) See Chapters 3 and 4

Swan LW (1967) Alpine and aeolian regions of the world. In: Wright HE Jr, Osburn WH (eds) Arctic and alpine environments. Indiana University Press, Bloomington, IN, pp 29–54

Zwinger AH, Willard BE (1972) Land above trees: a guide to American alpine tundra. Harper & Row, New York, 448 pp

Appendix References

Abell R, Thieme ML, Revenga C et al (2008) Freshwater ecoregions of the world: a new map of biogeographic units for freshwater biodiversity conservation. Bioscience 58(5):403–414

Bailey RG (1989) Explanatory supplement to ecoregions map of the continents. Environ Conserv 16:307–309. With separate map at 1:30,000,000

Bailey RG (1995) Description of the ecoregions of the United States. 2nd edn rev and expanded (1st edn. 1980). Miscellaneous publication no. 1391 (rev).

USDA Forest Service, Washington, DC, 108 pp With separate map at 1:7,500,000

Bailey RG (1996) Ecosystem geography. Springer, New York, 204 pp 2 pl. in pocket

Bailey RG (1998) Ecoregions: the ecosystem geography of the oceans and continents. Springer, New York, 176 pp 2 pl. in pocket

Bailey RG (2004) Identifying ecoregion boundaries. Environ Manage 34(suppl 1):S14–S26

Bailey RG, Zoltai SC, Wiken EB (1985) Ecological regionalization in Canada and the United States. Geoforum 116(3):265–275

Biasutti R (1962) Il paesaggio terrestre, 2nd edn. Unione Tipografico, Torino, 586 pp

Blasi C, Capotorti G, Smiraglia D et al (2010) The ecoregions of Italy. Ministry of Environment, Land and Sea Protection, Rome, Italy, 20 pp Available at http://www.minambiente.it/export/sites/default/archivio/biblioteca/protezione_natura/ecoregioni_italia_eng.pdf

Brown DE, Reichenbacher F, Franson SE (1998) A classification of North American biotic communities. University of Utah Press, Salt Lake City, UT, 141 pp

Cathey HM (1990) USDA plant hardiness zone map. USDA miscellaneous publication no. 1475. U.S. Department of Agriculture, Washington, DC

Chen SG, Morimoto Y (2009) Topographic watersheds as a framework for the new Japanese regional administrative units for ecosystem management. J Jpn Soc Revegetation Technol 34(1):287–290

Clements G (1916) Plant succession: an analysis of the development of vegetation. Publication 242. Carnegie Institution of Washington, Washington, DC, 512 pp

Clements G, Shelford VE (1939) Bioecology. Wiley, New York, 425 pp

Cleveland C (2011) Ecoregions (collection). Encyclopedia of Earth. Available at http://www.eoearth.org/view/article/151949/

Comer P, Gaber-Langendoen R, Evans SG et al (2003) Ecological systems of the United States: a working classification of U.S. terrestrial ecosystems. NatureServe, Arlington, VA, 83 pp Available at http://www.natureserve.org/library/usEcologicalsystems.pdf

Commission for Environmental Cooperation (2006) Ecological regions of North America, level III, map scale 1:10,000,000. Available at http://www.epa.gov/wed/pages/ecoregions/na_eco.htm

Commonwealth of Australia (2012) Interim biogeographic regionalization of Australia (map). Available at http://www.environment.gov.au/parks/nrs/science/bioregion-framework/ibra/index.html

Dasmann RG (1972) Towards a system for classifying natural regions of the world and their representation by national parks and reserves. Biol Conserv 4:247–255

Dice LR (1943) The biotic provinces of North America. University of Michigan Press, Ann Arbor, 78 pp

Dokuchaev VV (1899) On the theory of natural zones. Sochineniya (Collected Works) vol 6. Academy of Sciences of the USSR, Moscow-Leningrad, 1951

Ecoregion Working Group (1989) Ecoclimatic regions of Canada, first approximation. Ecological land classification series no. 23. Environment Canada, Ottawa, 119 pp With separate map at 1:7,500,000

Elder J (1994) The big picture: Sierra Club Critical Ecoregions Program. Sierra 1994:52–57

Ellis EC, Ramankutty N (2008) Putting people in the map: anthropogenic biomes of the world. Front Ecol Environ 8(6):439–447

European Environment Agency (2002) DMEER: digital map of European Ecological Regions. Available at http://www.eea.europa.eu/data-and-maps/figures/dmeer-digital-map-of-european-ecological-regions

Fenneman NM (1928) Physiographic divisions of the United States. Ann Assoc Am Geogr 18:261–353

Foreign Agricultural Organization (FAO) (2012) Global ecological zones for FAO forest reporting: 2010 update. Forest resources assessment working paper 179. Food and Agriculture organization of the United Nations, Rome, 42 pp

Fu B-J, Liu G-H, Lu Y-H, Chen L-D, Ma K-M (2004) Ecoregions and ecosystem management in China. Int J Sustain Dev World Ecol 11(4):397–409

Gallant AL, Binnian EF, Omernik JM, Shasby MB (1995) Ecoregions of Alaska. U.S. Geological Survey professional paper 1567. US Geological Survey, Washington, DC, 73 pp With separate map at 1:5,000,000

Hargrove WW, Luxmore RJ (1998) A clustering technique for the generation of customizable ecoregions. In: Proceedings ESRI Arc/INGO users conference. Available at http://research.esd.ornl.gov/~hnw/esri98

Harris SA (1973) Comments on the application of the Holdridge system for classification of world life zones as applied to Costa Rica. Arct Alp Res 5(3 pt 2):A187–A191

Herbertson AJ (1905) The major natural regions: an essay in systematic geography. Geogr J 25:300–312

Holdridge LR (1947) Determination of world plant formations from simple climatic data. Science 105:367–368

Host GE, Polzer PL, Mladenoff DJ, White MA, Crow TR (1996) A quantitative approach to developing regional ecosystem classifications. Ecol Appl 6:608–618

Huggett RJ (1995) Geoecology: an evolutionary approach. Routledge, London, 320 pp

Isachenko AG (1973) Principles of landscape science and physical-geographical regionalization (trans. from Russian by R.J Zatorski, edited by J.S. Massey). Melbourne University Press, Carlton, VIC, 311 pp

Jackson ST (2009) Alexander von Humboldt and the general physics of the Earth. Science 324:596–597

James PE (1959) A geography of man, 2nd edn. Ginn, Boston, 656 pp

Jepson P, Whittaker RJ (2001) Ecoregions in context: a critique with special reference to Indonesia. Conserv Biol 16(1):42–57

Loveland TR (ed) (2004) Special issue: ecoregions for environmental management. Environ Manag 34 (suppl 1):S1–S148

Lugo AE, Brown SL, Dodson R et al (1999) The Holdridge life zones of the conterminous United States in relation to ecosystem mapping. J Biogeogr 26:1025–1038

Mackey BG, Berry SL, Brown T (2008) Reconciling approaches to biogeographic regionalization: a systematic and generic framework examined with a case study of the Australian continent. J Biogeogr 35:213–229

Merriam CH (1898) Life zones and crop zones of the United States. Bulletin division biological survey 10. U.S. Department of Agriculture, Washington, DC, pp 1–79

Milanova EV, Kushlin AV (1993) World map of present-day landscapes: an explanatory note. Prepared by Moscow State University and the United Nations Environment Programme. Scale 1:15,000,000

Nowacki G, Spencer P, Fleming M, Brock T, Jorgenson T (2002) Unified ecoregions of Alaska: 2001. U.S. Geological survey open-file report 02–297 (map)

Olson DM, Dinerstein E, Wikramanayake ED, Burgess ND et al (2001) Terrestrial ecoregions of the world: a new map of life on Earth. Bioscience 51(11):933–938

Olstad TA (2012) Understanding the science and art of ecoregionalization. Prof Geogr 64(2):303–308

Omernik JM (1987) Ecoregions of the conterminous United States (map supplement). Ann Assoc Am Geogr 77:118–125

Passarge S (1929) Die landschaftsgürtel der erde, natur und kultur. Ferdinand Hirt, Breslau, 144 pp

Rowe JS, Sheard JW (1981) Ecological land classification: a survey approach. Environ Manage 5:451–464

Sayre R, Comer P, Warner H, Cress J (2009) A new map of standardized terrestrial ecosystems of the conterminous United States. U.S. Geological Survey professional paper 1768, 17 pp

Sayre R, Comer P, Hak J et al (2013) A new map of standardized terrestrial ecosystems of Africa. Association of American Geographers, Washington, DC, 24 pp

Schultz J (1995) The ecozones of the world: the ecological divisions of the geosphere (trans. from German by I. and D. Jordan). Springer, Heidelberg, 449 pp

Schwabe A, Kratochwil A (2011) Classification of biogeographical and ecological phenomena. In: Millington A, Blumier M, Schickhoff U (eds) Sage handbook of biogeography. Sage, London, pp 75–98

Simons H (2001) FRA 2000. Global ecological zoning for the Global Forest Resources Assessment 2000. Forest resources assessment working paper 56. FAO, Rome, 211 pp

Snelder T, Lehmann A, Lamouroux N, Leathwick J, Allenbach K (2009) Strong influence of variable treatment on the performance of numerically defined ecological regions. Environ Manage 44(4):658–670

Spalding MD, Gox HE, Allen GR et al (2007) Marine ecoregions of the world: a bioregionalization of coastal and shelf areas. Bioscience 57(7):573–583

Spalding MD, Agostini VN, Rice J, Grant SM (2012) Pelagic provinces of the world: a biogeographic classification of the world's surface pelagic waters. Oceans Coast Manage 60:19–30

Strahler AN, Strahler AH (1989) Elements of physical geography, 4th edn. Wiley, New York, 562 pp

Tansley AG (1935) The use and misuse of vegetation terms and concepts. Ecology 16:284–307

Trewartha GT, Robinson AH, Hammond EH (1967) Physical elements of geography, 5th edn. McGraw-Hill, New York, 527 pp

Tricart J, Cailleux A (1972) Introduction to climatic geomorphology (trans. from French by C.J. Kiewiet de Jonge). St. Martin's Press, New York, 274 pp

Troll C, Paffen KH (1964) Karte der Jahreszeiten-Klimate der Erde. Erdkunde 18:5–28

Udvardy MDF (1975) A classification of the biogeographical provinces of the world. Occasional paper no. 18. International Union for Conservation of Nature and Natural Resources, Morges, Switzerland, 48 pp

Walter H (1984) Vegetation of the earth and ecological systems of the geo-biosphere (3rd revised and enlarged edn) [trans. from German by Owen Muise]. Springer, Berlin, 318 pp

Walter H, Box E (1976) Global classification of natural terrestrial ecosystems. Vegetatio 32:75–81

Weigelt P, Jetz W, Kreft H (2013) Bioclimatic and physical characterization of the world's islands. Proc Natl Acad Sci U S A 110(38):15307–15312

Wiken EB (1986) Terrestrial ecozones of Canada. Ecological land classification series no. 19. Environment Canada, Hull, QC, 26 pp + map

Woodward SL (2009) Introduction to biomes. Greenwood Press, Westport, CT, 164 pp

Wu S, Yang Q, Zheng D (2003a) Delineation of eco-geographic regional system of China. J Geogr Sci 13 (3):309–315

Wu S, Yang Q, Zheng D (2003b) Comparative study of eco-geographic regional system between China and USA [in Chinese]. Acta Geograph Sin 58 (5):686–694

Glossary References

Holdridge LR (1947) Determination of world plant formations from simple climatic data. Science 105:367–368

Merriam CH (1898) Life zones and crop zones of the United States. Bulletin division biological survey 10. U.S. Department of Agriculture, Washington, DC, pp 1–79

Walace AR (1876) The geographical distribution of animals. With a study of the relations of living and extinct faunas as elucidating the past changes of the earth's surface, 2 vols. Macmillan, London, p 503, p 607

Walter H (1984) Vegetation of the earth and the ecological systems of the geo-biosphere (trans. from German by O. Muise) (3rd edn). Springer, Berlin, 318 pp

Maps

Plate 1 (left) Ecoregions of the Oceans

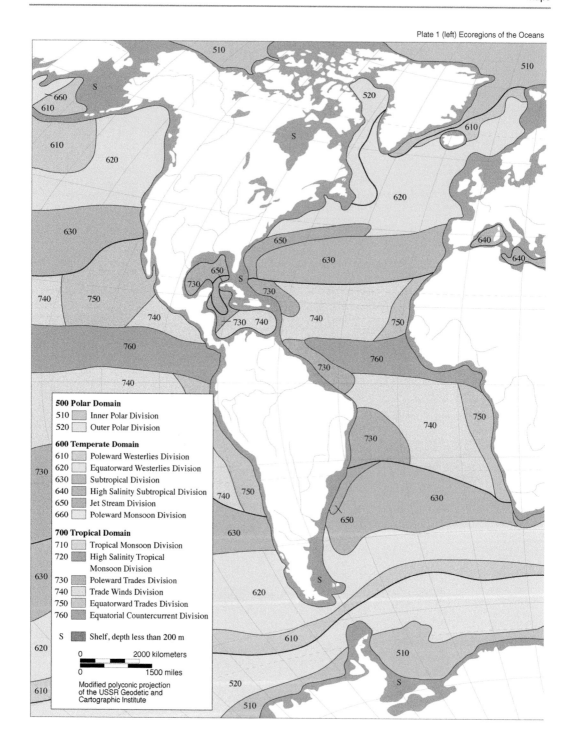

500 Polar Domain
510 ▢ Inner Polar Division
520 ▢ Outer Polar Division

600 Temperate Domain
610 ▢ Poleward Westerlies Division
620 ▢ Equatorward Westerlies Division
630 ▢ Subtropical Division
640 ▢ High Salinity Subtropical Division
650 ▢ Jet Stream Division
660 ▢ Poleward Monsoon Division

700 Tropical Domain
710 ▢ Tropical Monsoon Division
720 ▢ High Salinity Tropical
            Monsoon Division
730 ▢ Poleward Trades Division
740 ▢ Trade Winds Division
750 ▢ Equatorward Trades Division
760 ▢ Equatorial Countercurrent Division

S ▢ Shelf, depth less than 200 m

0   2000 kilometers
0   1500 miles

Modified polyconic projection
of the USSR Geodetic and
Cartographic Institute

Plate 1 (right) Ecoregions of the Oceans

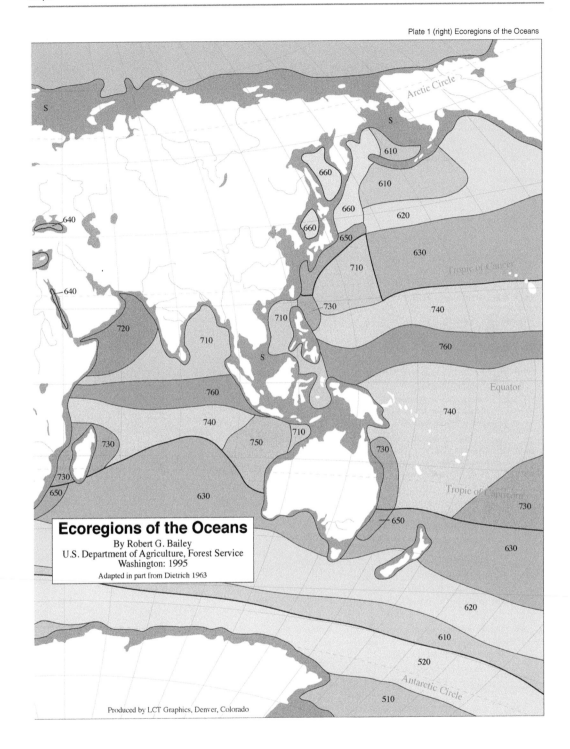

Ecoregions of the Oceans
By Robert G. Bailey
U.S. Department of Agriculture, Forest Service
Washington: 1995
Adapted in part from Dietrich 1963

Produced by LCT Graphics, Denver, Colorado

Plate 2 (left) Ecoregions

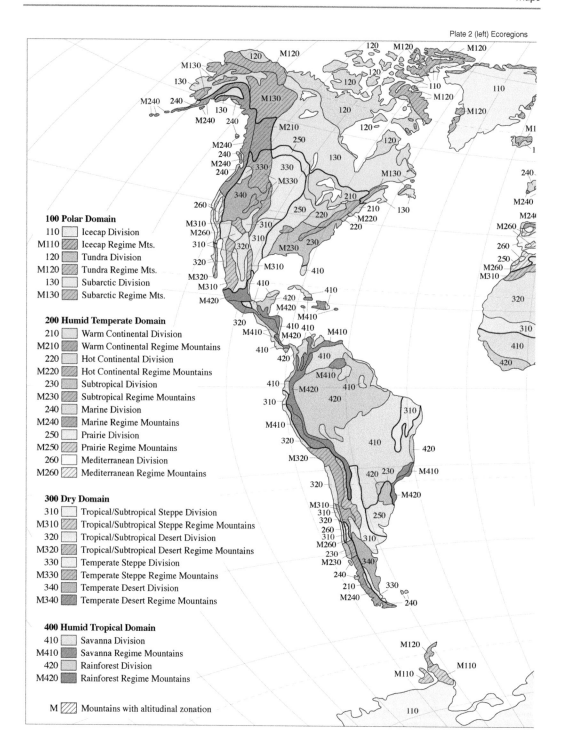

100 Polar Domain
110   Icecap Division
M110   Icecap Regime Mts.
120   Tundra Division
M120   Tundra Regime Mts.
130   Subarctic Division
M130   Subarctic Regime Mts.

200 Humid Temperate Domain
210   Warm Continental Division
M210   Warm Continental Regime Mountains
220   Hot Continental Division
M220   Hot Continental Regime Mountains
230   Subtropical Division
M230   Subtropical Regime Mountains
240   Marine Division
M240   Marine Regime Mountains
250   Prairie Division
M250   Prairie Regime Mountains
260   Mediterranean Division
M260   Mediterranean Regime Mountains

300 Dry Domain
310   Tropical/Subtropical Steppe Division
M310   Tropical/Subtropical Steppe Regime Mountains
320   Tropical/Subtropical Desert Division
M320   Tropical/Subtropical Desert Regime Mountains
330   Temperate Steppe Division
M330   Temperate Steppe Regime Mountains
340   Temperate Desert Division
M340   Temperate Desert Regime Mountains

400 Humid Tropical Domain
410   Savanna Division
M410   Savanna Regime Mountains
420   Rainforest Division
M420   Rainforest Regime Mountains

M   Mountains with altitudinal zonation

Plate 2 (right) Ecoregions

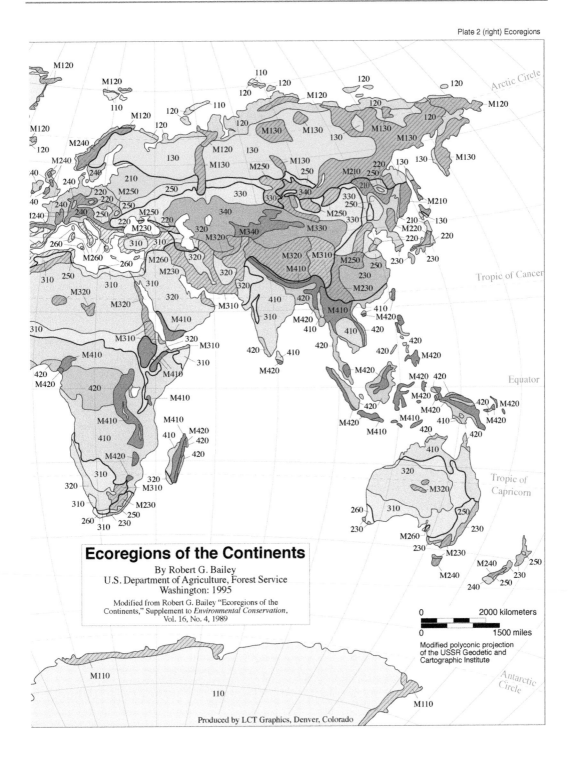

Ecoregions of the Continents

By Robert G. Bailey
U.S. Department of Agriculture, Forest Service
Washington: 1995

Modified from Robert G. Bailey "Ecoregions of the
Continents," Supplement to *Environmental Conservation*,
Vol. 16, No. 4, 1989

0   2000 kilometers

0   1500 miles

Modified polyconic projection
of the USSR Geodetic and
Cartographic Institute

Produced by LCT Graphics, Denver, Colorado

About the Author

Robert G. Bailey (b. 1939) received his PhD in geography from the University of California, Los Angeles (1971). A geographer with the U. S. Forest Service, Rocky Mountain Research Station, he was leader of the agency's Ecosystem Management Analysis Center for many years. He has four decades of experience working with the theory and practice of ecosystem classification and mapping, with applications in slope stability, land capability, inventory and monitoring, ecosystem management, climate change, and sustainability. His other books include *Ecosystem Geography: From Ecoregions to Sites* (Springer, 2nd ed, 2009) and *Ecoregion-Based Design for Sustainability* (Springer 2002).

Index

R.G. Bailey, *Ecoregions*, DOI 10.1007/978-1-4939-0524-9, © Springer Science+Media, LLC 2014